Bourbaki

A Secret Society of Mathematicians

Bourbaki

A Secret Society of Mathematicians

Maurice Mashaal

Translated by
Anna Pierrehumbert

American Mathematical Society

Original edition
Bourbaki:
Une société secrète de mathématiciens.
© Éditions Pour la Science - Paris, 2002   ISBN: 2-84245-046-9

Translated from the French by Anna Pierrehumbert.
English text © 2006 by the American Mathematical Society.

2000 *Mathematics Subject Classification.* Primary 01A70, 01A60

Front cover: Clockwise from the top left-hand corner: André Weil, Jean-Pierre Serre, Jean Dieudonné, Henri Cartan, Claude Chevalley, General Charles Bourbaki, Alexandre Grothendiecke, and Laurent Schwartz.

The poem by Arnold Seiken on page 168 was originally published in the *Mathematical Intelligencer,* vol. **17**, No. 2, p. 11, 1995. © 1995 with the kind permission of Springer Science and Business Media.

For additional information and updates on this book, visit
www.ams.org/bookpages/bourbaki

Library of Congress Cataloging-in-Publication Data
Mashaal, Maurice.
[Bourbaki English]
Bourbaki : a secret society of mathematicians / Maurice Mashaal.
p.cm.
Translation of: Bourbaki: une société secrète de mathématiciens.
Includes bibliographical references.
ISBN 0-8218-3967-5 (alk. paper)
1. Bourbaki, Nicolas. 2. Mathematicians—France—Biography. 3. Mathematics—History—20th century. I. Title.

QA29.B692M3713 2006
510.92′244—dc22   2006040704

∞ The paper used in this book is acid-free and falls within the guidelines established to ensure permanence and durability.
Visit the AMS home page at `http://www.ams.org/`
10 9 8 7 6 5 4 3 2 1   11 10 09 08 07 06

Contents

Preface 2

1. A Group Forms 4
2. The Story of a Name 22
3. Young Turks against Stubborn Priests 38
4. Bourbaki's Éléments de Mathématique 50
5. Towards Axioms and Structures 70
6. A Snapshot of Bourbaki's Work: Filters 86
7. The Bourbaki Seminar 96
8. Subtle and Austere Schoolboys 104
9. "For the Honor of the Human Spirit"? 116
10. New Math in the Classroom 134
11. An Immortal Mathematician? 146

Acknowledgements 154

Bibliography 155

Photo Credits 159

Appendix 162

Preface

Let's say it straightaway: Nicolas Bourbaki is not an individual but a group of mathematicians, almost all French. Formed in 1935, this colorful group consists of a dozen members and remains active today. Despite being unknown to most of the general public, these mathematicians changed the face of mathematics during the sixties and seventies.

Bourbaki neither invented revolutionary techniques nor proved grandiose theorems—and neither did it try to do so. What the group did bring, primarily through its imposing treatise *Éléments de mathématique*, was a new vision of mathematics, a profound reorganization and clarification of its components, lucid terminology and notation, and a distinctive style. It seduced many mathematicians, to such an extent that Bourbaki's philosophy has pervaded the international mathematics community. This has increased the influence of French mathematics throughout the world.

It was not only *Éléments de mathématique* that contributed to Bourbaki's fame. The exceptional quality of the group's members also contributed fundamentally to the quality of the group. André Weil, a key figure in Bourbaki since its creation, was one of the greatest mathematicians of the century. The other mathematicians who helped found the group, including Henri Cartan and Claude Chevalley, were also of international stature. More recent members include prestigious figures of mathematics such as Laurent Schwartz, Alexandre Grothendieck, and Jean-Pierre Serre. The members of Bourbaki all carried out individual mathematical work, earning them a range of high international honors. Several of them, notably Claude Chevalley, Laurent Schwartz, Alexandre Grothendieck, and Roger Godement, also committed much of their time to philosophy and politics. The group's philosophy also found its way into the New Math revolution of the seventies. While Bourbaki regretted this extension of their work, like Antigone they watched in fear as their views broke free from the group and started to lead their own existence.

In addition, Bourbaki constructed an extensive folklore around the group, building on the group's secrecy, its name, its humor and schoolboy pranks, its structure and method of working. Bourbaki is indebted to this folkore for much of the group's success. Yet the use of the word "success" must be qualified. Bourbaki did not complete its mission, and the evolution of mathematics will surely prevent it from ever being completed. Bourbaki and its philosophy had some unwelcome idiosyncracies that their critics readily denounced. And the future survival of this multi-brained mathematician is questionable.

Bourbaki's story is one of shadows and dazzling light. I hope I have shown both equally in this book.

—Maurice Mashaal

1

A Group Forms

On December 10, 1934, a handful of young mathematicians gathered in a cafe in the Latin Quarter of Paris. Their goal: to write a textbook on analysis. This is the beginning of a legendary enterprise that would change mathematics forever.

"André Weil and I were both at the University of Strasbourg in 1934. I often talked with him about the course on differential and integral calculus that I was teaching. At that time, a degree in mathematics was composed of certificates in general physics, differential and integral calculus, and classical mechanics. In other words, there was only one certificate in mathematics. [...] Therefore, I had to put as much as possible into what there was. I often wondered about the best way to teach this course because the existing textbooks were not satisfactory, especially when it came to multiple integrals and Stokes' Theorem. I discussed my concerns several times with André Weil. One beautiful day he told me, 'I've had it, we need to fix this for good. We need to write a good textbook on analysis. Then you'll stop complaining!'" This is how Henri Cartan described the origins of Bourbaki to Marian Schmidt in 1982, and André Weil confirms it in his *Souvenirs d'apprentissage* (translated as *The Apprenticeship of a Mathematician*), published in 1991: "One winter's day, in late 1934, I came up with a terrific idea for putting an end to my friend's persistent interrogation. I told him [Henri Cartan], 'There are five or six of us teaching the same course in various universities. Let's get together, fix the problems, and after that I'll finally be free from your questions.' I didn't realize that Bourbaki was born at this moment."

It's impossible to guarantee that human memory accurately recalls events from more than fifty years in the past. Yet even though this conversation from 1934 only gives a hint of the profound reasons for forming the group and of what would happen later, these two quotations do summarize the birth of the Bourbaki group.

The University of Strasbourg, where André Weil and Henri Cartan taught at the beginning of their careers.

The prehistory of Bourbaki's unique mathematical adventure began in the 1920s at l'École Normale Supérieure in Paris. Almost all the future Bourbaki members would study in this institution, a fact that remains true today. The ENS, as it is usually abbreviated, is a member of the Grandes Écoles, the elitist system of institutions of higher education so particular to France. Created in 1794, the ENS originally focused on producing professors of secondary education (including high school and the preparatory classes for entry into the Grandes Écoles). By the end of the nineteenth century, however, the Normaliens (as students of the ENS are called) increasingly turned to jobs in higher education and research. The ENS is composed of two faculties: humanities (with about thirty students per year in the 1920s) and sciences (with about twenty students per year). Admission is based on a highly selective exam following two or three years of preparatory classes. In the 1920s, the admitted students stayed at the ENS for three years. The first two were devoted to university classes and the third was devoted to preparing for the final examinations for teacher certification.

The entrance of l'École Normale Supérieure in 1924.

Bourbaki's Cradle, l'École Normale Supérieure

L'École Normale Supérieure has produced numerous authors, intellectuals, and politicians, including Raymond Aron, Jean-Paul Sartre, and Georges Pompidou. In the sciences as well, it is the ENS (along with l'École Polytechnique) that produces most of the elite of French research. At the beginning of the nineteenth century, most French mathematicians had studied at the École Polytechnique; by the end of that century, most had studied at the ENS. To cite a few examples, Gaston Darboux, Émile Picard, Paul Painlevé, Jacques Hadamard, Élie Cartan, René Baire, and Henri Lebesgue were all Normaliens; however, Henri Poincaré was a Polytechnicien.

It is at the prestigious ENS on the rue d'Ulm that the five future founding members of Bourbaki, Henri Cartan, Claude Chevalley, Jean Delsarte, Jean Dieudonné, and André Weil, met and became friends. Delsarte and Weil arrived in 1922, Cartan in 1923, Dieudonné in 1924, and Chevalley in 1926. These mathematicians took part in the adventures of Bourbaki from the beginning in 1934 and didn't leave until age forced them to. However, they were not the only mathematicians who participated in founding the group. Other mathematicians of the same generation were involved in Bourbaki's first steps to various degrees.

After the founding plenary conference held in July 1935, where the collective pseudonym Bourbaki was adopted, the nine official members of the group were Jean Coulomb, Charles Ehresmann, Szolem Mandelbrojt (the uncle of Benoit Mandelbrot, the father of fractals), René de Possel, and the five founding members already named. Of these nine, only Mandelbrojt was not a Normalien. Born in Poland, Mandelbrojt had completed his doctorate in Paris and had been teaching at the University of Clermont-Ferrand since 1929. Coulomb, who was more of a geophysicist than a mathematician, left the group relatively early, in 1937.

These young mathematicians, each at most thirty years old, held their first work meeting on Monday, December 10, 1934. Initially, their project was to write an analysis textbook to replace the unsatisfactory exisiting books—Édouard Goursat's book in particular—for use in French higher education. The meeting was held over lunch at the cafe A. Capoulade in the Latin Quarter of Paris, near the Pantheon. This cafe, which was located at 63 boulevard Saint-Michel on the corner with rue Soufflot, has since been replaced by one of the many American-style fast-food havens. Gathered around a basement table were Henri Cartan, Claude Chevalley, Jean Delsarte, Jean Dieudonné, René de Possel, and André Weil. Almost all of these

The students of "l'École" in 1924:
1. Henri Cartan,
2. Georges Canguilhem,
3. Jean-Paul Sartre,
4. Jean Dieudonné,
5. Raymond Aron,
6. René de Possel,
7. Charles Ehresmann,
8. Paul Nizan,
9. Louis Néel.

mathematicians held positions at univerisities outside of Paris: Cartan and Weil were at Strasbourg, Delsarte at Nancy, Dieudonné at Rennes, and de Possel at Clermont-Ferrand. They were in Paris for the Julia Seminar, held at 4:30 pm at the newly founded Henri Poincaré Institute on the second and fourth Mondays of each month of the academic year.

What did they say during this first meeting? We can answer this thanks to the work of Liliane Beaulieu, a mathematical historian from Quebec, who had access to the minutes of Bourbaki's meetings. For Weil, who already seemed to have precise ideas about what he and his friends should set out to accomplish together, it was necessary to "establish the content of the differential and integral calculus certificate for twenty-five years by writing a treatise on analysis" and that this treatise would be "as modern as possible." Weil anticipated that the publisher would be Hermann, where he knew the director, Enrique Freymann. Delsarte strongly supported the idea that the book, covering a wide range of material, would be written collectively. Animated discussion about the size of the project, the

The cafe A. Capoulade, where our secret society of mathematicians gathered in 1934.

schedule of publication, and the organization of the work followed. Cartan proposed that the work would be at most 1000 or 1200 pages; Delsarte proposed that the first volumes would be published in at most six months; Weil proposed that, after a few preliminary meetings, the group would create subcommittees in charge of designing each section of the book, and that the group would decide on a precise and final plan during a plenary meeting during the following summer vacation. They also started to discuss the nature and content of the future book, although of course one meeting was not enough to answer all their questions.

The Committee for Writing a Treatise on Analysis, as the group called itself at the time, met ten times between December 1934 and May 1935—about once every two weeks and, like the first meeting, on Mondays at the cafe A. Capoulade. During the second meeting, the group decided that the Committee would include at most nine members. Although this rule was only enforced briefly following the birth of Bourbaki, the number of members never increased to more than about a dozen. In January 1935, even before the founding plenary conference that would take place that summer, Paul Dubreil, Jean Leray, and Szolem Mandelbrojt joined the group; however, Dubreil and Leray quit before the summer. Jean Coulomb replaced Dubreil in April, and Charles Ehresmann replaced Leray in the summer. Arrivals

At the first Bourbaki conference, held in Besse-en-Chandesse in 1935. Standing (left to right): Henri Cartan, René de Possel, Jean Dieudonné, André Weil, and a university lab technician. Seated (left to right): Mirlès (a guinea pig, or potential future Bourbaki), Claude Chevalley, Szolem Mandelbrojt.

and departures, which were marked by no ceremony, remained common throughout the history of Bourbaki, and the composition of the group varied continually.

A Modest Project Turns Ambitious

As for the project's nature and content, the initial intention of creating a book used for teaching differential and integral calculus quickly turned into a much more ambitious task. Already during the second meeting, Weil stated that "we must write a book that can be used by anyone: by researchers (both students and professors), by future teachers, by physicists, and by engineers." To accomplish this, the book would need to provide the readers with a collection of mathematical tools "as robust and universal as possible," and the group would need to develop a detailed outline for the book before deciding what tools to present. At the same time, the book would need to simplify the tools as much as possible; that is, to determine the real substance of these tools and to present the most general, and therefore universal, versions of them. This is not the case of most textbooks, where one of the weaknesses is often that fundamental theorems "are presented with a pretty incredible restraint: far

The small village of Besse-en-Chandesse, where the first Bourbaki conference was held.

more hypotheses than necessary are used."

Forming a detailed outline of the book turned out to be a long process, certainly much longer and arduous than the members of Bourbaki initially anticipated. The planning took several years and was very productive. Gradually, the group's extensive reflection and lively discussions led to a new vision of mathematics, a modern way of teaching it and even of doing it. The volumes of the book presented this new vision and had a great influence on the mathematical community, both in France and throughout the world.

The first complete outline was established during the founding plenary conference, which took place from July 10 through July 17, 1935, in Besse-en-Chandesse, a small village about twenty-five miles from Clermont-Ferrand. The main subjects to be included in the book (such as functions of real and complex variables, integrals, differential equations and partial differential equations, and special functions) were approximately the same as in earlier textbooks on analysis. In addition, the group planned to add a small number of more abstract and innovative chapters providing basic concepts of abstract algebra, set theory, and topology, which the group deemed necessary for the coherent presentation of the main material. The entire work was expected to cover 3200 pages, which is three times more than the Committee had anticipated during its first meeting, held just a few months earlier!

Szolem Mandelbrojt, Claude Chevalley, René de Possel, and André Weil (from left to right) during the 1935 conference.

Bourbaki, as the group was named at the Besse-en-Chandesse conference (the first name Nicolas was added later), planned to complete a first draft of the book within a year. While this goal was far from achieved, it did set the group's plans in motion. Of course there were still debates, questions, hesitation, and backtracking, all of which resulted in modifications and refinements of the outline for the book. Over the course of the following years, the "abstract packet" (Bourbaki's name for the sections dealing with abstract algebra and topology, which were to serve as a foundation for the rest of the book) grew relentlessly, while the other chapters were delayed and ~~and~~ even deemphasized. Initially conceived as auxiliary to the other chapters and intended to be as short as possible, the abstract packet became one of the main parts of the project. In fact, Bourbaki's project had become so broad and ambitious that the phrase "treatise on analysis" no longer applied. Thus it was under the title *Éléments de mathématique*—the resemblance to Euclid's *Elements* is no coincidence—that Bourbaki's first volume appeared in 1939–1940. The first volume to appear was *Fascicule des résultats de théorie des ensembles*, a summary of set theoretical results presented without proof.

During World War II, despite the scattering of the Bourbaki members, the group managed to publish three more volumes of *Éléments de mathématique*. Numerous other volumes followed in the decades

René de Possel
The portaits ringed in blue show the future members of Bourbaki at their entrance to l'École Normale Supérieure.

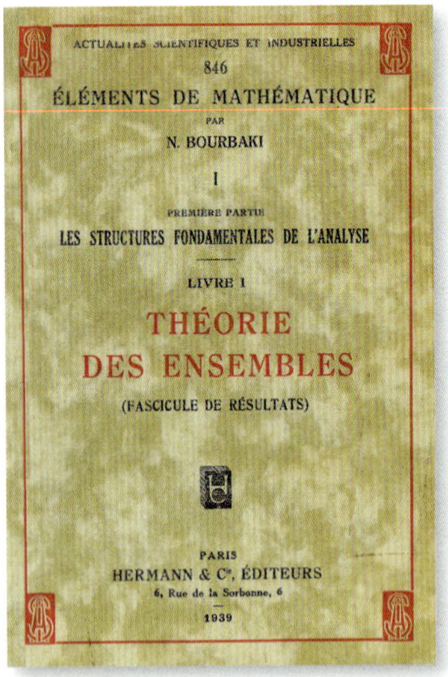

Bourbaki's first work: a summary of results in set theory.

between 1940 and 1970, with the production rate slowing down markedly after this. The most recent volume was published in 1998, while the one before that had been published fifteen years earlier, in 1983.

Éléments de mathématique (which currently contains about 7000 pages dense with definitions, axioms, lemmas, corollaries, and theorems)—and especially a few particular volumes—was a worldwide success and made Bourbaki famous. But the success and fame of the group stem equally from the distinctive way in which the members lived and worked. Such a huge and influential treatise succeeded not only because of the mathematical talent of its authors, but also because of the group's enthusiasm, friendship, schoolboy solidarity, and belief in a common goal.

The Bourbaki group is devoid of any hierarchy. All decisions must be unanimous, and while there is no official voting, anyone can veto a proposal. This system particularly applies to decisions about the treatise. The final version must be accepted by every member, which requires, in most cases, many years of work. The writing process itself is distinctive as well. The group gives the task of writing the rough draft of a given section to one or two members. Once the section is written, it is read out loud to all the members, after which the other members mercilessly criticize the draft. The group then assigns the task of rewriting the section to a different member. The process continues until it converges—sometimes wearily—to a manuscript that the members unanimously accept as ready for publication.

However, the lack of hierarchy doesn't imply that all the members have equal weight in the group. Some members put more into the group; some have more influence. André Weil, who can be considered as the group's first leader, is *primum inter pares*, the first among equals. He was also the target of jokes and mockeries less often than the other members. Even Jean Dieudonné, a vocal member who worked intensely for the group, was aware of Weil's role. Henri Cartan tells how "one day, Dieudonné said [metaphorically], 'I won't drink my *café au lait* before Weil.'" In the more recent history of Bourbaki, key members of the group include Jean-Pierre Serre, Michel Demazure, Pierre Cartier, and Jean-Louis Verdier.

Laurent Schwartz

André Weil, Armand Borel, and Jean-Pierre Serre on the terrace of the hotel in Pelvoux-le-Poët (in the Alps), July 1951.

Conferences in the Country

Bourbaki holds three conferences each year to keep track of progress and make decisions about the future. These meetings usually take place in the countryside, in calm and agreeable places. In recent years, each conference lasted one week; in the past, when summer vacations were longer, each conference lasted two weeks. These conferences, gathering a dozen mathematicians working seven or eight hours a day, seemed to proceed amidst a cheerful hubbub. There were always many people talking loudly at the same time. Jokes rang out as frequently as insults, which were delivered more or less purposefully but always quickly forgotten. The group enjoyed an animated atmosphere of comradeship away from work as well, and

André and Eveline Weil in 1939. André Weil (1906–1998), the brother of the philosopher Simone Weil, was one of the great mathematicians of the twentieth century. He played a central role in Bourbaki, even after he left France in 1941.

many people who spotted the group during one of their conferences called them "a bunch of lunatics"!

On the other hand, this bunch of lunatics behaved (and behaves) much more discreetly outside their conferences. One of the most striking peculiarities of the Bourbaki enterprise is that no outside person is supposed to know who the current members are, what the group is working on, or when and where the conferences take place. During a series of interviews with former Bourbaki members, led by the reporter Michèle Chouchan and aired on France-Culture in 1988, Laurent Schwartz, an early Bourbaki recruit, said that "whenever I was asked if I was a member of Bourbaki, I had to say no. If I wasn't a member, I was speaking the truth, and if I was a member, I was required to say that I was not." Also, at the Bourbaki secretariat, housed in an office at the École Normale Supérieure, one is politely told that Bourbaki doesn't help newspapers or accept interviews, and that it "neither confirms nor denies any information circulating about the group." In short, it is difficult to obtain direct information. Only former members agree to break the silence.

There are several reasons for this tradition (or mania?) of secrecy. According to Bourbaki, the group preserves its secrecy to preserve the collective nature of their enterprise. Bourbaki writes its books as a collective effort, and no member must be allowed to put himself ahead of others, be it for scientific merit or for the collection of royalties received from sales of the group's books. Of course, the secrecy also allows the group to work productively without being disturbed; the fact that the group was most secretive during its golden years from 1950 to 1970 supports this view. This secrecy may also protect the group's members from influential scientists who were skeptical or hostile towards the project, and such people did exist from the very beginning. Withholding the names of the members also strengthens the authority of the group's treatise: the contents of the books appear as the expression of a consensus, with any disagreements among the members of the group remaining invisible. Finally, the secrecy served a social function: strengthening the cohesion of the group and helping to create the myth surrounding it. This is not the least among the charms of Bourbaki.

It appears that Bourbaki was initially less secretive than it became in later years. An example of this lack of secrecy is a letter written on November 17, 1936 by Mandelbrojt, Delsarte, Cartan, Weil, Dieudonné, de Possel, Coulomb, Chevalley, and Ehresmann to the physicist Jean Perrin (at the time the State Undersecretary for Scientific Research). This letter, a grant request, summarizes what Bourbaki was and describes the financial difficulties of the project, which at the time was far from producing royalties. This is what Bourbaki wrote:

"Dear Mr. Perrin:

"Perhaps you have already heard that the mathematicians named above are devoting a large amount of their time to writing a Treatise on Mathematical Analysis, which—we hope—will be the basis of analysis education for the next decade.

"We have adopted a new method of collaboration. We have not restricted ourselves by distributing the task of writing distinct pieces of the whole work; instead, each chapter is assigned to one of us only after we have discussed and prepared it at great length, and the resulting draft is again discussed in detail after every member has studied it. Each chapter is revised at least once, and often several times. Through this method, we hope to obtain a truly collective work, one with a profoundly unified character.

"It is clear that the method we have chosen is not a lazy one, and that it requires numerous meetings and frequent travel. Furthermore, much of our time is spent on reproducing and distributing various drafts and manuscripts. For the past two years, we ourselves have taken responsibility for financing these activities. Now that the State officially sponsors scientific research, we thought that it might be able to help us. It is this financial help, Mr. Perrin,

André Weil and Armand Borel in 1955. André Weil was formidably caustic. To a mathematician who asked him, "Can I ask you a stupid question?" André Weil replied, "You just did..."

Samuel Eilenberg (1913–1998), one of the fathers of category theory, at a Bourbaki conference. Eilenberg was one of the few non-French members of the group.

that we request of you with all respect.

"Following is a summary of the funds we are requesting. Seven of us live outside of Paris and we hold at least four meetings each year; allowing on average 250 francs per person per meeting to cover travel and living expenses yields a total of 7000 francs. In addition, we estimate that 3000 francs are needed for various expenses, including correspondance, office supplies, typing, duplication, and especially the payment of assistants who copy formulas by hand onto the duplicated manuscripts, a service which is as vital as it is expensive.

"A grant of 10000 francs per year for four or five years would allow us to complete our project successfully.

"Mr. Perrin, we hope that you will accept our profound deference and respectful admiration."

The letter closed with the signature of Mandelbrojt, who was the oldest member of the group; the pseudonym Bourbaki was never mentioned. The grant was awarded for a year and subsequently renewed.

A Guinea Pig Must Prove Himself

Bourbaki quickly gained new members through a unique recruitment procedure. Bourbaki often invited one or two additional people to the group's conferences. In some cases, these were "guinea pigs," potential future recruits who were being tested. Once a promising young mathematician had been spotted, Dieudonné explained at a conference in Romania in 1968, "the procedure was to invite him to a conference as a 'guinea pig,' it's a traditional method. You all know about laboratory guinea pigs, who are exposed to all sorts of viruses. Well, it's the same sort of thing, the poor guy is thrown into the Bourbaki discussion sessions, and not only must he understand, he must also participate. If he's silent, he simply won't be invited again."

If he is recruited, it means that other members must leave, since Bourbaki hardly ever grew beyond a dozen members. Sometimes, departures stem from more or less significant disagreements about the group's goals and methods of working. For example, Paul Dubreil left even before the founding plenary conference of 1935, in part because of unavailability (his wife had been hired at Rennes, while he was at Nancy) and in part because, as he confided to Liliane Beaulieu, he disliked the disorderly discussions and preferred working with one or two people on precise problems than to working in a large group. Jean Leray, an eminent French mathematician, also left early because his proposals had been impertinently shot down and also because, according to Beaulieu, he opposed the group's principle of starting from the very basics of mathematics. Also, as in any social group, there were discords of a more personal nature. This appears to be the case of René de Possel: in 1937, his wife Éveline became Éveline Weil.

The Association of Collaborators of Nicolas Bourbaki

To manage the group's various practical issues, and especially its financial affairs, Bourbaki made itself official by creating a not-for-profit organization. This "Association of Collaborators of Nicolas Bourbaki" was registered on August 20, 1952 at the prefecture in Nancy. Its headquarters was Jean Delsarte's house, at 4 rue de l'Oratoire in Nancy. In 1966, the headquarters was moved to Jean-Pierre Serre's address in Paris, and it has been at l'École Normale Supérieure, at 45 rue d'Ulm in Paris, since 1972.

One should differentiate the Bourbaki group from the Assocation of Collaborators of Nicolas Bourbaki. Although the members are the same, the association is only an administrative body dealing with the outside world. Therefore, the statutes of the association don't mention the specific rules about secrecy, unanimity, retirement at fifty, the recruitment process, and so forth.

The first board of directors consisted of Jean Delsarte (president), Henri Cartan (vice-president), Jean Dieudonné (secretary), Jean-Pierre Serre (treasurer), Roger Godement, Laurent Schwartz, Jacques Dixmier, and Pierre Samuel. In 1995, the general assembly of October 20 was presided by Bernard Tessier, with the other members present being Arnaud Beauville, Gérard Ben Arous, Daniel Bennequin, Patrick Gérard, Guy Henniart, Pierre Julg, Olivier Mathieu, Joseph Oesterlé, Marc Rosso, Georges Skandalis, and Jean-Christophe Yoccoz.

DECLARATION d' EXISTENCE
-:-:-:-:-

Le soussigné ; déclare , conformément à la loi du I Juillet I90I , qu'une association ayant pour titre " ASSOCIATION des COLLABORATEURS de Nicolas BOURBAKI "

Pour objet : Toutes études , recherches et travaux en vue de l'avancement des sciences mathématiques -

La publication et la communication des travaux des membres de l'association dans toutes revues et à toutes Académies ou Sociétés Savantes , publiés sous le pseudonyme de N.BOURBAKI , notamment de l'ouvrage intitulé " Eléments de Mathématiques par N.BOURBAKI .

L'organisation de conférences et congrès ; la participation des délégués de l' Association à des manifestations similaires .

La prise de contatt avec toutes personnalités , Ecoles , Universités Françaises ou étrangères poursuivan des recherches mathématiques .

A été fondée le 2 Juillet I952.

Son siège social est à Nancy , rue de l'Oratoire N°4

Elle est administrée par un conseil composé de :

I°) Monsieur le Doyen DELSARTE (Jean Frédéric Auguste), Professeur à la Faculté des Sciences de l'Université de Nancy demeurant à Nancy rue de L'Oratoire N°4 PRESIDENT

2°) Monsieur CARTAN (Henri Paul)Professeur à la Sorbonne demeurant à Paris (I4°) Boulevard Dourdan N°95 VICE PRESIDENT

3°) M. DIEUDONNE (Jean Alexandre Eugène)Professeur à la Faculté des Sciences de Nancy , demeurant à Nancy rue Saint Miche N°26 SECRETAIRE

4°) Monsieur SERRE (Jean Pierre) chargé de recherches au Centre National de Recherches Scientifiques , demeurant à Paris (I0°) Boulevard de la Chapelle N°39 TRESORIER

5°) Monsieur GODEMENT (Roger Jean Henri) Professeur à la Faculté des Sciences de l'Université de Nancy , demeurant à Nancy MEMBRE

6°) Monsieur SCHWARTZ (Laurent) Professeur à la Sorbonne demeurant à Nancy Cours Léopold N°30 MEMBRE

7°) Monsieur DIXMIER (Jacques) Professeur à la Faculté des Sciences de Dijon , demeurant à Paris (I3°) rue Le Brun N°I5 MEMBRE

8°) Et Monsieur SAMUEL (Pierre)Professeur à la Faculté des Sciences de Clermont , demeurant à Royat avenue Antoine Phelut MemBRE

Fait à Nancy le 30 Août 1952

Le Président du Conseil d' Administration .

Delsarte

Jean-Pierre Serre

Retirement at Fifty

Another reason for departure from Bourbaki was age, since members must retire at fifty. This rule was initiated by Weil when the founding members were approaching this age. At the end of Dieudonné's birthday lunch during the summer conference of 1956, held at Sallières-les-Bains in the department of Drôme, Henri Cartan read a letter from Weil (who had been living in the United States since 1947 and only came to one in three conferences), which proposed a "gradual disappearance of the founding members."

According to Weil, two considerations motivated his proposal. The first was that "the number of conference participants is sometimes too high to allow fruitful work," and the second was that "the founding members are 'more equal than others,' and thus the more recent members 'don't feel obliged to take on full responsibility' in group discussions." But surely an additional motivating factor was the widely held belief that a mathematician is most brilliant and productive in his youth. As Dieudonné said in 1968, "[...] a mathematician above fifty can still be a very good mathematician, still very productive, but he rarely can adapt to new ideas, ideas developed by people 25 or 30 years younger than him. And an enterprise like Bourbaki would like to be eternal [...]."

Thus most of the founding members left around 1956-1958, and the fifty year rule was retained in the following years. In some sense, Bourbaki is eternally young! However, the retirement of a member does not mean that he is entirely cut off from the active members, since for example the retired members continue to receive *La Tribu* ("The Tribe"), the group's internal newsletter describing the proceedings of its conferences.

In Bourbaki's sixty-five years of existence, the group has involved about forty members, almost all of which were French and students of the ENS. Among the exceptions (who all spoke French) are Samuel Eilenberg, an American of Polish origin who founded category theory with Saunders MacLane around 1942 and collaborated with Bourbaki for fifteen years until 1966; Armand Borel, a Swiss living in the United States and a member for twenty years until 1973; and Serge Lang, an American of French origin.

In addition to the mathematicians already cited, notable members of Bourbaki include Arnaud Beauville (b. 1947), Claude Chabauty (1910-1990), Alain Connes (b. 1947), Jacques Dixmier (b. 1924), Adrien Douady (b. 1935), Roger Godement (b. 1921), Alexandre Grothendieck (b. 1928), Jean-Louis Koszul (b. 1921), Charles Pisot (1909-1984), Pierre Samuel, and Bernard Teissier. It would be difficult to form a complete list of the members of Bourbaki, and even harder to determine their exact dates of collaboration with the group. These dates can only be approximated roughly from the fact that mathematicians joined at about 25 years and left before 50.

Jean Dieudonné

Great Talent, Great Minds

Bourbaki earned fame and influence in part because of the scientific strength of its members. All of the members of Bourbaki were, or still are, very good or even excellent mathematicians with great productivity outside of the Bourbaki projects. Many members won a Fields Medal, an international prize whose prestige is equivalent to that of a Nobel Prize which does not exist for mathematics. The Fields medallists participating in Bourbaki are Laurent Schwartz (1950), Jean-Pierre Serre (1954), Alexandre Grothendieck (1966), Alain Connes (1982), and Jean-Christophe Yoccoz (1994). Meanwhile, many excellent French mathematicians were not members of Bourbaki, such as René Thom (the creator of catastrophe theory and a Fields medallist in 1958), Marcel Berger, André Lichnerowicz, and Jean Leray (who was a member only very briefly following the birth of Bourbaki).

Several members of Bourbaki are also important and famous outside of mathematics. Laurent Schwartz and Alexandre Grothendieck are just two examples. To the public outside of mathematics, Schwartz is known not for his 1944 creation of distribution theory (of great importance in the theory of partial differential equations) but for his rich political activity dealing with Algeria, Vietnam, Soviet mathematicians, and the French university system. He recounts these activities extremely well in his autobiography, *Un mathématicien aux prises avec le siècle* (published in English as *A Mathematician Grappling with His Century*). As for Alexandre Grothendieck, this genius mathematician produced highly abstract work and revolutionized algebraic geometry, a vast subject based on solutions to systems of polynomial equations and connected to many domains of mathematics, including number theory. Unfortunately for the scientific community—which he criticized with increasing vehemence as the years wore on—this unequalled mathematician brutally cut himself off from mathematics around 1970 in favor of campaigning for the environment and living an isolated existence in southern France.

A bunch of lunatics, highly secretive and bearing a strange name, formed of colorful characters who were also brilliant mathematicians;

an original *modus operandi*; a process of regeneration that guarantees eternal youth; a massive treatise whose 7000 pages brought fame and influence—all this is Bourbaki. But why the name "Bourbaki"? Why so much work devoted to a book initially destined simply for teaching differential and integral calculus? What is so noteworthy about their treatise anyways? In short, why all the fuss?

Jean Delsarte (1903–1968)

Jean Delsarte was born on October 19, 1903 in Fourmies, a small town in Northern France, into a Catholic family where the father was a retired director of textile factories. As André Weil writes in a biographical note, "Delsarte remained faithful to his religious beliefs, which he combined with rare open-mindedness. This certainly played an important role in his behavior and way of thinking." After studying in Fourmies, Tréport, and Paris, Jean Delsarte obtained a scholarship in 1915 to study at the Lycée Corneille in Rouen, where he succeeded brilliantly. He graduated in 1921 and entered l'École Normale Supérieure in 1922. Having earned his teacher certification in 1925, Delsarte served in the military and in 1926 became a boarder at the Fondation Thiers, an institution created in Paris in 1893 to accomodate young scholars for three years while they carried out research. He stayed there only a year (which did give him the time to write his doctoral thesis) before he was hired as a junior lecturer in the Science Faculty at Nancy. He thus moved to Nancy, where he would remain for much of his career. He married Thérèse Sutter, a childhood friend, in 1929.

Delsarte applied his energy to making Nancy into an active center of mathematics. He set up good connections with Strasbourg, where his friends Henri Cartan and André Weil had been teaching since 1931 and 1933 respectively. An "Eastern branch" of the French Mathematical Society gradually formed, as Delsarte used his influence to strengthen mathematics at Nancy, in particular by the hiring of Paul Dubreil, Jean Leray, and Jean Dieudonné. "Starting in 1934," André Weil writes, "Delsarte played a significant role in forming the team of collaborators of Nicolas Bourbaki." In 1937, he was the main figure in the "Medal Wars," a revolt against the system of monetary and honorary rewards created by the physicist Jean Perrin, who had founded the CNRS (Centre National de la Recherche Scientifique, the French scientific research institute) and was serving as the first undersecretary of state for scientific research at the time.

In September 1939, Jean Delsarte was mobilized; as a captain, he commanded an audio reconaissance battery. He managed to lead the safe retreat of his unit from Alsace to Nîmes, where he was discharged. He taught in Grenoble for the 1940–1941 academic year to substitute for the professor Favard, who had been taken prisoner. He then returned secretly to Nancy, where he continued his teaching and research,

At the July 1951 conference: Jean-Pierre Serre and Jean Dieudonné (standing); Jacques Dixmier, Laurent Schwartz, and Jean Delsarte (seated).

which dealt with the expansion of functions into power series—he was a virtuoso when it came to calculation—as well as number theory and some problems in mathematical physics. In late 1942, Delsarte formed a group to discuss the reform of scientific research in France. After the war, he also participated in the Langevin-Wallon committee for teaching reform, from which he returned rather disgusted because he didn't think it addressed any real issues. Starting in 1947, Delsarte accepted various invitations to spend time at universities abroad; among others, he visited the Institute for Advanced Study in Princeton, the University of São Paulo, Mexico, and Bombay. Meanwhile, he also made Nancy into a renowned center of mathematics, with brilliant minds like Dieudonné, Laurent Schwartz, and Roger Godement. However, this success was only temporary, as the attraction of Paris eventually drew students and scientists to the capital. He left in 1962 to direct the France-Japan House in Tokyo and returned to Nancy in 1965. His fragile health (he suffered from ophthalmological problems and severe diabetes) soon started to take its toll. Then came the French student uprisings of May 1968, "which he didn't withstand," Weil writes. Delsarte was a notable, he was "dean Delsarte" and was used to deference. "Delsarte couldn't understand that the chaos was necessary to create (perhaps) a new society and consequently a new university system." Did this weaken him even more? Jean Delsarte died of a heart attack on November 28, 1968.

2

The Story of a Name

Why "Bourbaki"? Originating in École Normale Supérieure folklore, this pseudonym would become a central part of the myth surrounding our group of young French mathematicians.

On July 16, 1935 at Besse-en-Chandesse in Auvergne, during the first conference held by the founders of Bourbaki, our mathematicians decided to take a break from an unproductive discussion about important results on analytic functions. They went to Pavin Lake, about three miles from Besse-en-Chandesse. There, according to one of the group's internal documents (cited by Liliane Beaulieu), "A few members had the courage to dive stark naked into the waves while yelling 'Bourbaki!' hundreds of times." While this evokes the sacred incantations during a rite of initiation into a cult, this act was in fact merely the tribulations of a joyful bunch of young scientists rather than those of a dubious religious sect.

The Association Bourbaki and rue Bourbaki in Pau are named after the General, not the multi-brained mathematician.

The Real General Bourbaki (1816–1897)

Born into a family of Greek origin, Charles Bourbaki was educated at l'École Spéciale Militaire, France's most prestigious military academy. He participated in the campaign in Africa from 1836 to 1854 and became the colonel of a regiment of zouaves in 1851. He then fought in the Crimean War from 1854 to 1856 as a member of the Eastern army. He became a brigadier general in 1854, and was promoted to division general after returning from an expedition to Algeria. He participated in the campaign in Italy (1859–60) and was an inspector general for the infantry as well as a division commander from 1860 to 1869. He became the Emperor's aide-de-camp in July 1869 and the commander in chief of the Imperial Guard one year later. During the Franco-Prussian War of 1870–1871, he participated in several battles in the East (Borny, Rezonville, Amanvillers, Sainte-Barbe) before becoming the commander of the First Army in September 1870. He was victorious in the battle of Villersexel in January 1871 but suffered a terrible defeat at Héricourt a week later. This forced him to retreat through Besançon and Switzerland, where his troops disarmed (and where he tried to commit suicide). He was made a division commander and then the military governor in Lyon and was placed on reserve in 1879.

But who was this "Bourbaki"? A character out of a Tintin comic book? No. Merely a fictitious product of imagination? No. In fact, the pseudonym chosen by our group of brilliant mathematicians has its origins in the pranks played by Normaliens, the students at the École Normale Supérieure. In 1923, the third-year student Raoul Husson decided to play a prank on the first-year students. He posted an announcement that a Professor Holmgren would be coming to give a lecture, and that all the first-year students should attend. The result? Here is what André Weil wrote in his *Souvenirs d'apprentissage*: "Wearing a false beard and speaking with an unplaceable accent, he introduced himself to the students and proceeded to give a lecture. In the lecture, which dealt with a bit of classical function theory, he advanced by tiny degrees to apparently extravagant heights, concluding with 'Bourbaki's Theorem.' This lecture became legendary, and added to the legend was the assertion that one of the students claimed to understand the entire lecture."

Raoul Husson found the name Bourbaki in French military history. In the army of Napoleon III was a general named Charles Denis Sauter Bourbaki, who played an important part in the Franco-Prussian War

At the Bourbaki conference in Murols, August 1954: Laurent Schwartz, Henri Cartan, and André Weil.

of 1870–1871. This Charles Bourbaki was forced to lead his troops to Switzerland to escape from Prussian forces. During World War II, these actions were coincidentally replicated by a battery including a member of the Bourbaki group. Jean Delsarte, one of the founders of Bourbaki, was mobilized at the beginning of the war and became a captain in charge of an audio reconnaissance battery. He was forced to lead the retreat of his unit from Alsace in Northeast France to Languedoc in Southern France via the Jura Mountains and the Alps. Imagine Delsarte's surprise when, one day while crossing the Jura Mountains along the Swiss border, he heard one of the soldiers proclaim, "We are the army of Bourbaki"!

Now the only missing link in the chain from the army general to our group of mathematicians is how it occurred to Raoul Husson to use the name of General Charles Bourbaki in his lecture. At the time, the war of 1870–1871 was not so far in the past, and this general, although not very well-known today, had not been forgotten. Furthermore, Bourbaki's troops included some Normaliens. Thus, while a general anti-military atmosphere prevailed at the ENS during the twenties, Bourbaki's name remained familiar to the Normaliens. However, Henri Cartan gave Marian Schmidt a somewhat different explanation of the name's origin. According to Cartan, "We think that when he [Raoul Husson] was a student at l'École Normale Supérieure, he took courses on military subjects, just as everyone had to. These were being taught by a captain who gave presentations in such a dogmatic tone that in

André Weil in Aligarh with Vijayaraghavan and two students, 1931.

the school's end-of-the-year 'revue' (unfortunately, a tradition that has since been lost), Raoul Husson gave a lecture consisting of a sequence of theorems, all bearing the names of generals! There is the origin of the name Bourbaki used for our treatise."

Schoolboy Prank or Literary Allusion?

Sterling K. Berberian, a mathematician at the University of Texas, suggested another possible reason for the group's adoption of the name Bourbaki. In 1980, Berberian remarked in The Mathematical Intelligencer (a non-technical quarterly magazine providing entertainment for mathematicians) that the satirist and anarchist Octave Mirbeau used the name Bourbaki in his 1900 work, *Le journal d'une femme de chambre* (*The Diary of a Chambermaid*). One of the characters, a retired captain who brags about eating anything, tames a ferret named Kléber and then a hedgehog named Bourbaki. This hedgehog is "an extraordinary and intelligent animal, a prankster, who eats anything! [...] It's incredible...it's impossible to fill him up...he's like me...he'll eat anything!" Berberian suggests that the group of mathematicians adopted the name Bourbaki because it reflects the ambition and mathematicial gourmandise of the treatise the group intended to write. However, none of the members of Bourbaki have alluded to Mirabeau's omnivorous hedgehog. As for Raoul Husson, we do not know whether he read Mirbeau, and even if he did, whether the book provided the name he used in his prank.

Perhaps André Weil—who loved languages and literature and who was the main force in creating the group and its name—knew the work, and if this was the case it certainly would have added to the name's appeal. But he never alluded to the hedgehog either. Be that as it may, Weil must have liked Raoul Husson's prank because he described it to his young mathematician friend D. Kosambi while living in India in 1930–1932. According to Liliane Beaulieu, Kosambi was having a power struggle with one of his colleagues. Weil suggested that he write an article relating to the imaginary work of an imaginary Russian mathematician named Bourbaki. Since the colleague in question would obviously be unaware of this mathematician or his work, such an article would deflate his ego.

Kosambi did just that. He published a farcical article called "On a generalization of the second theorem of Bourbaki" in the bulletin of the Academy of Sciences of the Agra and Oudh Allahabad provinces. This text, dated 1931–1932 and sprinkled throughout with wordplay most likely suggested by Weil, discusses the work of "an unknown Russian mathematician," D. Bourbaki, who "was poisoned during the Revolution." Among others, Kosambi thanked Weil for making him aware of the "important work" of this Bourbaki. It is in this way that the name Bourbaki made its first appearance in a mathematical publication.

About three years later, this name became the pseudonym of the handful of mathematicians who decided to publish a modern treatise on analysis. One of the advantages of using a pseudonym, other than the folklore it evokes and the myth it builds, is that it provides a simple solution to the problem of how to sign the treatise. For the founders

THÉORIE DE LA MESURE. — *Sur un théorème de Carathéodory et la mesure dans les espaces topologiques.* Note (1) de M. **Nicolas Bourbaki**, présentée par M. Élie Cartan.

La théorie moderne de la mesure et de l'intégration (2) conduit à donner le nom de *mesure*, dans un ensemble quelconque (qui prend alors le nom d'*espace mesuré*), à toute fonction d'ensemble μE satisfaisant aux axiomes suivants :

I. *A tout ensemble* E *de points de l'espace correspond un nombre* μE *tel que* $0 \leqq \mu E \leqq +\infty$; *si* E *est vide*, $\mu E = 0$.

II. *Si* E *est contenu dans la réunion d'ensembles* E_ν *en nombre fini ou dénombrable, on a*

$$\mu E \leqq \sum_\nu \mu E_\nu.$$

(1) Séance du 18 novembre 1935.

(2) Voir, par exemple, S. Saks, *Théorie de l'intégrale* (*Monog. Matem.*, 3, Varsovie, 1933). Cf. de Possel, *C. R. du Séminaire mathématique de M. Julia*, 2, 1934-1935, p. 1-19 et *Comptes rendus*, 201, 1935, p. 579.

An extract from Nicolas Bourbaki's article in *Comptes Rendus de l'Académie des Sciences* (1935).
(For English translation, see appendix.)

Émile Picard was the secretary of l'Académie des Sciences in 1935.

of Bourbaki, it was out of the question to sign the treatise with a long list of the names or initials of the contributors, who would not remain the same from one year to the next.

Although the surname Bourbaki was adopted during the founding conference of July 1935, the first name Nicolas was not added until later. At first, the group decided to use the initial N in front of the surname, since this was the letter traditionally used in course announcements if the teacher had not yet been chosen. The question of adding a first name first arose in Autumn 1935, when the group decided to officially establish the existence of Bourbaki. To do this, the members of Bourbaki wanted to publish a short article signed with their pseudonym in the *Comptes Rendus de l'Académie des Sciences*.

Now, the French Académie des Sciences requires that every author submitting work to its journal provide a short biographical sketch. Moreover, the work and the biographical sketch must be given to a member of the Académie, who then passes them on to the Académie if he deems them acceptable. The biographical sketch was the subject of a discussion among several members of Bourbaki, which Weil recalls in his *Souvenirs d'apprentissage*: [...] to do this, we needed to give Bourbaki a first name. Éveline, my future wife, was present at the discussion, and she became Bourbaki's godmother; she baptised him Nicolas. We also needed a member of the Académie to present his biography; we were sure that Émile Picard, the secretary of the Académie, would have an apoplectic fit if he found out what we were up to. I took responsibility for writing the biography and sending it to Élie Cartan with a letter of support."

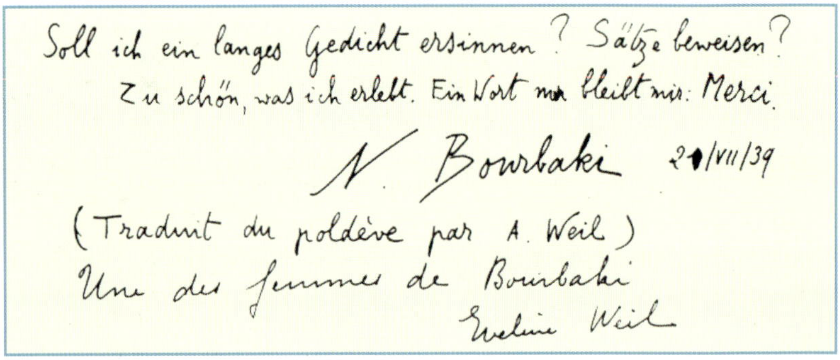
Soll ich ein langes Gedicht ersinnen? Sätze beweisen?
Zu schön, was ich erlebt. Ein Wort nur bleibt mir: Merci.
N. Bourbaki 21/VII/39
(Traduit du poldève par A. Weil)
Une des femmes de Bourbaki
Eveline Weil

Bourbaki's autograph in 1939, from the guest book of Rolf Nevalinna in Korkee, Finland.
(For English translation, see appendix.)

Élie Cartan Sponsors a Wacky Mathematician

Élie Cartan, a member of the Académie des Sciences, was the father of Henri Cartan (a member of Bourbaki), and he was among the very few mathematicians of his generation to be respected by the Bourbakis. Although he didn't participate in Bourbaki's activities, Élie Cartan knew about the group's project and held a certain understanding for these young Turks and their project.

Élie Cartan (1869–1951) was one of the few non-Bourbaki mathematicians respected by the group (left). Jean Leray (1906–1998) left Bourbaki soon after joining because he didn't agree with the group's philosophy (right).

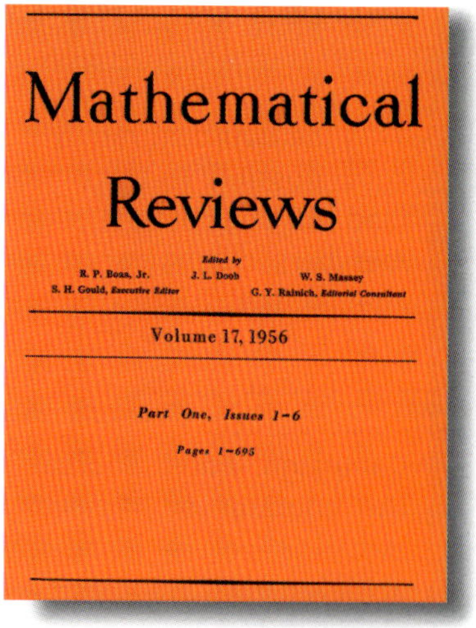

The cover page of the 1956 *Mathematical Reviews* naming the editorial board, which included Ralph P. Boas.

Thus André Weil wrote an article called *Sur un théorème de Carathéodory et la mesure dans les espaces topologiques* that discussed a certain issue in the theory of integration. In his letter to Élie Cartan, Weil wrote that "I am sending you the attached, an article that Mr. Bourbaki gave me to pass on to you, for publication in the *Comptes Rendus*. I'm sure you'll recall that Mr. Bourbaki is the former professor of the Royal University of Besse-en-Poldévie whom I met some time ago at a cafe where he spends most of the day and even the night, having lost both his job and most of his fortune amid the troubles that caused the unfortunate Polvedian nation to disappear from Europe. Now he earns his living at the cafe by giving lessons in belote, the card game he plays so brilliantly. He claims not to think about mathematics anymore, but he was gladly willing to talk with me about a few important problems and even to let me take a look at some of his papers. I managed to persuade him to publish the attached article, which contains a highly useful result in the modern theory of integration. [...]"

Élie Cartan presented the article and the curious biography of its author to the other members of the Académie des Sciences on November 18, 1935 at the end of lunch, before the Académie convened for its session. According to the account Henri Cartan gave to Liliane Beaulieu, the lunch had been accompanied by plentiful wine, and the bizarre story did not unduly surprise the members of the Académie. Of course, the fact that the article was impeccable made the story behind it somewhat irrelevant. Whether or not the Académie members (many of whom were former Normaliens, and thus accustomed to pranks) were fooled or not, Nicolas Bourbaki's article was accepted and published before the year was out.

Thus Bourbaki was baptised as Nicolas in 1935 (although the volumes of the treatise *Éléments de mathématique* continued to be signed as N. Bourbaki for many years). The first name was chosen in part because our young mathematicians thought Bourbaki sounded Russian. However, they made a mistake: Nicolas Bourbaki was Greek!

An Unknown Relative Appears

Like a character in a picaresque novel, our hero discovered a miraculous relative. The discovery arrived one autumn morning in 1948. While Henri Cartan was eating breakfast, the telephone rang. His wife Nicole picked it up and, a few seconds later, returned to her husband saying, "It's Bourbaki on the phone." This wasn't a prank. On the other end of the line was Nicolaidis Bourbaki, at the time the commercial attaché to the ambassador of Greece in Paris and briefly an officer in the American army during the occupation of Germany.

Ralph Boas (1912–1992).

"Notes on the Life and Work of Nicolas Bourbaki"

"The Bourbaki family is of Cretan origin. It is possible to trace it back to the year 1089 (and the original family name Scordylis), when its members distinguished themselves in resisting against the Turks. Impressed by this family's bold exploits, the Turks called them the Vour bachi (that is, 'he who strikes first'), and the name stuck. During Napoleon's Egyptian Campaign, Sôter Bourbaki (1750–1820) served the emperor by breaking through the British blockade several times; in recognition of his achievements, Napoleon allowed Bourbaki's two eldest sons to be raised at the expense of the French State. The eldest son became a colonel in Napoleon's Grande Armée and was the father of the famous general of the Franco-Prussian War. Sôter Bourbaki's third son moved to Russia and then Romania and soon lost contact with his parents, who had stayed in Greece. Nicolas Bourbaki, who was born in Cucuteni (Moldova) in 1886, is descended from this youngest son. Numerous descendents of the various branches of the Bourbaki family still live in Greece as well as other parts of the world (with some currently living in Chicago); we are indebted to them for the historical facts mentioned above.

"After performing brilliantly in high school in his native country, Nicolas Bourbaki attended classes at the University of Kharkov before obtaining a fellowship to study in Paris with H. Poincaré and in Göttingen with D. Hilbert, who profoundly influenced his way of thinking. He defended his thesis at the University of Kharkov in 1910. This paper (of which the few copies printed were all destroyed in the German invasion of 1941) already shows the seeds of Bourbaki's future development. He became a privatdozent at the University of Dorpat in 1913 and married two year later. He and his wife would have one daughter, Betti, who married the lion-hunter H. Pétard in 1938. World War I interrupted Bourbaki's fruitful scientific work. He then found himself in the Caucasus during the Revolution of 1917, working in a research institute in Poldevia. Elected a member of the Royal Academy of this country, Bourbaki was keenly interested in the Poldevian people (related to the Ossetions, the Poldevians are one of the many races found in the Caucasus mountains). But the civil war forced Bourbaki to emigrate to Iran in 1920.

"This was the beginning of a dark period in Bourbaki's life, in which he spent several years as a refugee in search of a stable position that would allow him to live decently. He first worked with a French geophysical expedition in Iran and then spent some time as a professor at the Royal College of Zorngahr in India. He finally ended up in Paris, where no mathematician would acknowledge the originality of his ideas. He lived miserably for several years on precarious means.

"Nevertheless, it appears that some echos of his ideas penetrated scientific circles, although no one understood them well, and students tended to make fun of this 'unknown mathematician.' In 1923, his name was cited in a ritual prank at l'École Normale Supérieure, in which a so-called Professor Holmgren gave the new students a most fantastic lecture that ended with a wonderful (yet incomprehensible) theorem proved by Bourbaki.

"No more examples of his work appeared until 1930, when one of his Indian students, the well-known mathematician Kosambi, cited one of Bourbaki's important results in his work. But the second period of Bourbaki's scientific work didn't start until 1935. At this time, many French mathematicians of the younger generation had been struck by the emerging tendency towards specialization and scattering in contemporary mathematics. Lacking simple and solid directing principles, these young mathematicians feared that their discipline would crumble into a dusting of many subjects with no communication among them, a sort of mathematical Tower of Babel.

"Luckily, these students happened to meet Bourbaki and realized that Bourbaki's views agreed with theirs. They managed to persuade Bourbaki to work with them in publishing a great work that would present the most important elements of mathematics in a completely new way. This proved to be a longterm enterprise: 23

At the conference in Dieulefit, 1938 (left to right): Simone Weil, Charles Pisot, André Weil, Jean Dieudonné (sitting), Claude Chabauty, Charles Ehresmann, and Jean Delsarte.

volumes (of about 150 pages each) have already been published, while more are nearing completion. The final work will consist of about fifty volumes, which its authors hope will constitute our era's analog of Euclid's Elements (whence the treatise's name). It must be emphasized that this is not an encyclopedia. The theories to be presented were strictly chosen among existing mathematical theories with the guiding principle that they must 1) be useful in the widest possible range of cases and 2) be as general as possible, which relies on the theory being reduced to an abstract skeleton (however, abstraction is not used for its own sake, but rather with applications always kept in mind). All in all, the work is designed to teach young mathematicians as efficiently as possible how to work with the tools of their trade. The fruits of these labors are already clear, as the younger generations of mathematicians are assimilating fundamental mathematical theories with a speed that astounds their elders.

"The collaborators of Nicolas Bourbaki, who scorn fruitless discussions about 'priority' and believe that only teamwork could yield success in such an enterprise, chose to publish their Treatise under the name of their Master and to deliberately waive any acknowledgement for their individual work within the group.

"Twenty-five years of experience have proven beyond doubt that this method of working as a group is effective. For the first time in the history of the world, this shows that it is possible to make scientific discoveries in areas as abstract as mathematics by using the power of several minds joined through a long collaboration. All of Bourbaki's collaborators believe that this method will become very important in the future, as their Master has predicted already for a long time. But one must not imagine that such results can be achieved immediately. The fundamental principle of selflessness, the scorn of vain scientific glory, the measured fusion of the participant's intellectual personalities into a sort of higher collective mind: all these are absolutely necessary for such a group to function.

"This involves a sort of asceticism whose principles revealed themselves only gradually over the course of the group's experiences. The result of this constant interaction among such diverse minds is the 'spirit of Bourbaki,' which dramatically affirms its unity through its twenty-three volumes already published.

"The collaborators of Nicolas Bourbaki gather approximately three times per year for conferences of various lengths. The collaborators use these meetings to carefully examine and discuss the numerous drafts that each chapter must go through before it is deemed fit to publish. In the group's quest for the best presentation possible, for the presentation best reflecting the spirit of Bourbaki, each chapter passes through at least 3 or 4 consecutive 'states' (each usually written by different members). Some chapters have been drafts for more than twenty years.

"Although the group's fundamental rules prevent the names of the collaborators of Nicolas Bourbaki from being revealed, we can say that the composition of the group does not remain static. As mathematics is always evolving, it is necessary to keep the spirit of Bourbaki young. To this end, the group eagerly welcomes any worthy young mathematicians who have become interested in the enterprise. After attending the conferences as 'guinea pigs,' they are put to the test as 'chrysalises.' Once they have proven that they can participate effectively in the collaboration, they are finally accepted as full members. Currently, the ages of Bourbaki's collaborators range from 24 to 45 years. Moreover, some of the early members have left the group, as their interests wandered to other research or as the characteristic atmosphere of the group's discussions (where insults and curses fly freely) repelled or tired them. It is both expected and desirable that the oldest members would be invited to retire when their capacity to adopt new ideas wains, possibly slowing down the progress of the treatise, the group's most important activity. Only our Master himself is exempt from this rule, thanks to his exceptionally youthful spirit.

"Nicolas Bourbaki, who became somewhat of a misanthropist following his misfortunes, refuses to see anyone except the collaborators he has chosen for himself. This gave rise to the legend that he is merely a pseudonym, but anyone who has met him knows the strength and vitality of his extraordinary personality, to which even his collaborators are prone to ascribe somewhat mysterious powers. This is what causes, in the midst of discussions in which Bourbaki does not even take part, all the present members to simultaneously have sudden flashes of inspiration that show them the solutions to particularly thorny problems whose solutions had been elusive until then. In addition, while several of Bourbaki's most fierce opponents suffered terrible misfortune during the war, his collaborators all survived without the slightest harm, despite conditions where most risked their liberty or even their life. Is this a mere coincidence, or does the spirit of Bourbaki transcend its members?"

Henri Cartan and Nicolaidis Bourbaki decided to meet that very day at the Henri Poincaré Institute. According to Cartan, who told the story of this meeting to Marian Schmidt, "After showing me his papers to prove his identity, he explained that the news about a group of people taking the name Bourbaki had reached intellectual circles in Greece thanks to an article written by André Lichnerowicz for a magazine called *La Vie intellectuelle* ["Intellectual Life"]. Since the Bourbakis are an important family in Greece, Nicolaidis Bourbaki decided to find out [...] about the group of people who had permitted themselves to use the name without permission. This diplomat first visited the editor of the books by Bourbaki, namely Mr. Freymann [at éditions Hermann]. [...] In response to the diplomat's questions, Freymann simply replied, 'Go and see Henri Cartan.'"

Thus, Nicolaidis Bourbaki came to know the group of mathematicians who borrowed his name. Bonds between the diplomat and the group grew, and for several years Nicolaidis was a guest at dinners during the group's conferences. According to Henri Cartan, contact was broken upon the death of Nicolaidis' wife. As for the story of the Bourbaki family, the Bourbaki group incorporated it into the various wacky legends that the group propagated about itself.

The Notes on Nicolas Bourbaki

In the non-mathematical articles that the group wrote for amusement, Bourbaki liked to mix invention with truth. This is particularly the case for *Notice sur la vie et l'oeuvre de Nicolas Bourbaki* ("Notes on the Life and Work of Nicolas Bourbaki"), which was probably written around 1960 and which the American Judith Friedman reproduced in her 1977 thesis on Bourbaki. Most of this article is faithful to reality and is well worth reading (see pages 30–32).

In writing this article, the group was trying to make people believe that Nicolas Bourbaki was not a pseudonym but in fact a real mathematician, a "Master" with almost mysterious powers who attracted collaborators like a magnet. The group tried to propagate this assertion in other ways as well. For example, Bourbaki applied for individual membership to the American Mathematical Society in the fifties; however, the application was rejected, since most mathematicians knew that Bourbaki was actually the name of a group. But this didn't prevent Bourbaki from complaining! At around the same time, the group also complained to the American mathematician Ralph Boas about a short article he wrote on Bourbaki for an annual edition of the *Encyclopedia Britannica* in which he asserted that the name was a collective

The courtyard of l'École Normale Supérieure showing the Ernest Pond, named after the mathematician Ernest Poisson, on the right. The Bourbakis often held discussions around this pond.

pseudonym. Boas knew this from a reliable source, since he had met Weil in 1939, before Bourbaki had become so secretive. Nicolas Bourbaki immediately complained to the *Encyclopedia Britannica* and sent Boas an insulting letter that expressed astonishment that Boas could "dare to say that I don't exist." To get even, Bourbaki circulated a rumor that the mathematician Boas didn't exist, that B.O.A.S. was merely the acronym of the members of the editorial board of *Mathematical Reviews* (of which Boas was indeed a member).

Nicolas Bourbaki was more than just a first name and a last name. The mathematicians who invented him made him into a sort of mythical figure, a general of mathematical armies. They used it to play pranks on those outside the group. Moreover, they used him to serve an internal purpose by producing laughter in abundance and by providing the central element of the folklore surrounding the group.

André Weil (1906–1998)

Considered to be one of the greatest mathematicians of the century, André Weil was born on May 6, 1906 in Paris. His father, a doctor, was an Alsatian Jew and his mother was an Austrian Jew born in Russia. His sister Simone (1909–1943) would become an eminent philosopher. Most of Weil's elementary and secondary education was provided at home by excellent teachers in private lessons. He entered junior year at the Lycée Saint-Louis in 1918. After graduating, he prepared for the entrance exams into l'École Normale Supérieure and entered that school in 1922, at the age of just sixteen.

André Weil would become a world traveller. After being certified as a teacher in 1925, he spent a year in Rome studying with Vito Volterra on a modest fellowship. A Rockefeller fellowship allowed him to spend one year in Germany—spent at Göttingen, Berlin, and Frankfurt—where he met eminent mathematicians like Richard Courant, Emmy Noether, Karl Siegel, and Max Dehn. He interrupted this trip with a short visit to Stockholm to study with Gösta Mittag-Leffler (the mathematician said to have courted the same woman as Alfred Nobel, which caused mathematics to be excluded from the Nobel Prizes—but this, of course, may be a legend).

Upon returning to France, André Weil worked on his doctoral thesis *L'arithmétique sur les courbes algébriques*, which he defended in 1928 at the age of twenty-two. Next he travelled to India, where he was a professor at the Muslim University of Aligarh from 1930 to 1932; he tried to reform mathematics teaching there, but unfortunately did not really succeed. He then returned to France, first becoming a lecturer at the University of Marseille and then moving to the University of Strasbourg upon being urged to do so by his friend Henri Cartan. In 1937, he married Éveline, the ex-wife of René de Possel, one of the first members of Bourbaki. "My life, or at least the part of it worthy of that name [...] falls between May 6, 1906, the day I was born, and May 24, 1986, the day my wife and companion Éveline died," Weil writes in his autobiography. During the summer of 1939, André Weil and his wife took a half-touristic, half-professional voyage to Northern Europe, including Finland. He planned to emigrate to America from there if war broke out. A reserve officer like all students who entered l'École Normale Supérieure before 1923, Weil had decided to desert when faced with war. His reasons did not originate in pacifism; instead, influenced by Indian philosophy, André Weil believed that "there is no universal prescription for how everyone should behave. Each person holds his dharma within himself [...], and each person must determine his dharma as best he can. Gauguin's dharma was painting. Mine, as I saw it in 1938, seemed clear: it was to do mathematics as long as I was able. It would have been a sin to turn away from that." Regardless of the reaons, Weil's behavior met incomprehension and even enmity. The mathematician Jean Leray (who had been imprisoned and devoted his work in captivity to pure algebraic topology to make sure his results would not be useful to the Germans) particularly opposed Weil. This later prevented Weil from pursuing his career in France, and both Weil and his family regretted his exile. "Leray efficiently caused Weil to be excluded from the Sorbonne and the Collége de France. Weil remained at Chicago, and my generation lost a master!" writes Pierre Cartier, who was a member of Bourbaki from 1955 to 1983.

Weil found himself in Finland at the outbreak of World War II. In November 1939, a combination of circumstances led him to be suspected of spying for the Soviet Union. According to Weil, only a fortuitous conversation with the chief of police and the Finnish mathematician Rolf Nevanlinna, who knew him, saved him from the firing squad. Expelled from Finland, he arrived at le Havre only to be arrested and transferred to a military prison in Rouen. In May 1940, Weil was judged guilty of insubordination and avoided serving five years in prison by serving in the army before being demobilized in November. In 1941, he managed to reach the United States with his wife thanks to a rescue program for French scientists organized by the Rockefeller Foundation. He spent 1945 through 1947 as a professor at the University of São Paulo and then returned to the United States to become a professor at the University of Chicago. He stayed at Chicago for eleven years before accepting a position at the famed Institute of Advanced Study in Princeton. He retired in 1976 and died in Princeton on August 6, 1998.

Weil's mathematical work, dealing most notably with number theory and algebraic geometry, is vast. Despite his stand against the institution of prizes and medals in France, Weil earned numerous prestigious awards such as the Wolf Prize (in 1979) and the Kyoto Prize (in 1994). Even his doctoral thesis attracted attention. In addition, Weil proved the Riemann hypothesis for algebraic curves over finite fields. While trying to generalize this result to polynomials of several variables, Weil formulated a series of conjectures that led to numerous developments in algebraic geometry as they were being proved in the decade between 1963 and 1973. Another notable element of his work deals with the connections between the arithmetic of elliptic curves and that of modular forms. With the Japanese mathematicians Goro Shimura and Yukata Taniyama, he helped formulate a conjecture (called the Shimura-Taniyama-Weil Conjecture) that played a crucial role in the 1994 proof of Fermat's Last

André Weil with his wife Éveline.

André Weil teaching at the University of Chicago.

Theorem (for n an integer greater than 2, there exist no nonzero integers x, y, and z such that $x^n + y^n = z^n$) by the British mathematician Andrew Wiles. And this is just a small sampling of Weil's work.

It's also important to underline Weil's predominant role in Bourbaki. Weil was at the origin of this group and its peculiarities, and he was a central figure in the group until his retirement around 1956—all this despite the geographical distance that separated him from France.

Weil was not only interested in mathematics, and his education in the humanities was expansive. Although philosophy did not particularly interest him, Weil was fascinated by literature and languages. At fifteen, he already knew Greek, Latin, German, English, and some Sanskrit. Those who knew Weil willingly admit that he did have a difficult character, to say the least. His humor was caustic. Armand Borel writes that "André Weil had a sharp sense of humor and would quickly see the comic or ridiculous aspect of any situation or statement. This would lead to a biting, or even wounding, comment, or a spiritful, amusing one." Pierre Samuel, a member of Bourbaki from 1947 to 1971, said that "Weil was particularly harsh towards people who believed they were great mathematicians even though they weren't." According to Catherine Chevalley, whose father was a good friend of Weil, he was not mean but he hated flattery, especially when directed towards himself. Some of the aspects of André Weil's personality and humor show in *Souvenirs d'apprentissage* (published in English as *The Apprenticeship of a Mathematician*), his autobiography covering the years until he moved to Chicago in 1947.

3

Young Turks against Stubborn Priests

World War I extinguished a generation of French scientists. The older French mathematicians were dying, while the German algebraists worked with great vitality. This situation is what triggered the creation of Bourbaki.

The pet target of the first Bourbakis: Édouard Goursat's *Cours d'analyse.*

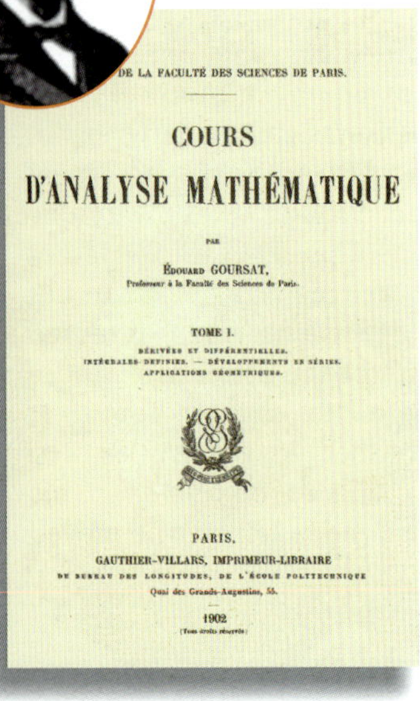

DE LA FACULTÉ DES SCIENCES DE PARIS.

COURS

D'ANALYSE MATHÉMATIQUE

PAR

ÉDOUARD GOURSAT,

Professeur à la Faculté des Sciences de Paris.

TOME I.

DÉRIVÉES ET DIFFÉRENTIELLES.

INTÉGRALES DÉFINIES. — DÉVELOPPEMENTS EN SÉRIES.

APPLICATIONS GÉOMÉTRIQUES.

PARIS,

GAUTHIER-VILLARS, IMPRIMEUR-LIBRAIRE

DU BUREAU DES LONGITUDES, DE L'ÉCOLE POLYTECHNIQUE

Quai des Grands-Augustins, 55.

1902

(Tous droits réservés)

Why did the Nicolas Bourbaki group form in the thirties? First, recall that the principal goal of this handful of young mathematicians was to write a treatise on analysis intended for use in university classrooms. This treatise was supposed to correct the flaws and insufficiencies of the existing textbooks, in particular those of *Cours d'analyse* by Édouard Goursat, the first edition of which was published in 1902 and which was widely used throughout France.

However, the inadequacy of the French analysis textbooks at the time was just the tip of the iceberg; in some ways, this inadequacy was merely the reflection of more profound problems in French mathematics. At the time, French mathematics was dwarfed by cutting-edge research in other countries, in particular by the fruitful German algebraists. The general situation of French mathematics at the beginning of the twentieth century allows us to understand not just the emergence of the Bourbaki group but also why the group's initial project quickly developed into a massive, decades-long project of revolutionizing mathematics.

Before describing the mathematical climate at the time of Bourbaki's creation, it would be useful to give an idea of how mathematics has developed across the centuries. Rather than recounting the entire history of mathematics, or even just a summary of it, we will just briefly mention a few key mathematical achievements throughout history.

Contrary to what one might expect, the knowledge already acquired in Antiquity—particularly in Greece, Egypt, Mesopotamia, India, and China—~~are~~ was far from negligible. The Pythagorean Theorem (in a right triangle, the square of the hypotenuse is the sum of the squares of the other two sides), for example, was discovered during this era. Although the concept of an equation had not been developed yet, geometric methods for solving quadratic equations ($ax^2 + bx + c = 0$, where a, b, and c are given and x is the unknown to determine) and systems of linear equations (for example, a system of two equations of the form $ax + by = c$, where x and y are unknown) were already known. The Ancient Greeks had discovered the incommensurability of the side and diagonal of a square, which is equivalent to the fact that $\sqrt{2}$ is not a rational number (a number of the form p/q, where p and q are integers), since the diagonal of a square with side length a is of length $a\sqrt{2}$. Around 300 B.C., Euclid wrote his *Elements*, an astonishingly modern work that presented the mathematical knowledge of the time concisely and relatively rigorously through a sequence of definitions, axioms, and theorems followed by their proofs. Other notable achievements of this era include the beginnings of planar trigonometry (by Hipparcus in the second century before Christ), the beginnings of spherical trigonometry (by Menelaus around 100 A.D.), and Diophantus's work on finding rational solutions to indeterminate systems of equations, which involve more unknowns than equations (around 350 A.D.).

During the Middle Ages, Western mathematics was struck by lethargy and subsequently regressed; meanwhile, the golden age of Arab mathematics started in the ninth century. Arab mathematics was founded on the discoveries by the Greeks and Babylonians, as well as the more recent discoveries by the Indians (who invented decimal notation for real numbers and the digit 0 some time before the seventh century). Arab mathematicians achieved numerous developments in arithmetic, algebra (a word of Arabic origin), and geometry. Among their developments and discoveries are decimal fractions, the binomial formula (which provides the expansion of $(a + b)^n$), a method for calculating roots of integers, polynomial arithmetic, calculations in combinatorics, and a method for solving cubic equations using conic sections.

Thales

Euclid

Pythagoras

The Renaissance: the Revival of Western Mathematics

Arab mathematics didn't start to reach the West until the eleventh century, when the first translations of Arab work were made. It was Leonardo of Pisa (approximately 1170–1240), also known as Fibonacci, who introduced Arabic digits in place of Roman numerals. However, European mathematics was not really revived until the sixteenth century, which saw the emergence of brilliant Italian algebraists

Leonardo of Pisa, nicknamed Fibonacci

including Niccolo Fontana (also known as Tartaglia), Girolamo Cardano, Ludovico Ferrari, and Rafael Bombelli. Their work includes the general solution of cubic and quadratic equations and the introduction of complex numbers (such as $2 + 3i$, where i is an "imaginary" number, the square root of -1 satisying $i^2 = -1$). It was also during the sixteenth century that François Viète developed a system of symbolic notation for algebra, the forerunner of the system in modern mathematics.

Mathematics entered a new era in the seventeeth century. John Napier (1550–1617) and Henry Briggs (1561–1630) invented logarithms. Girard Desargues (1591–1661) introduced some preliminary concepts of projective geometry, the study of the geometric properties of figures obtained by projection. Pierre de Fermat (1601–1665) and Blaise Pascal (1601–1665) gave birth to probability theory. Fermat and René Descartes (1596–1650) created analytic geometry, in which coordinates represent points and equations represent curves, creating an equivalence between geometric problems and algebraic calculations.

A major scientific advance of the era was Newton's and Leibniz's creation of differential and integral calculus in the 1670s. The central concept of differential calculus, which originated in the study of tangents to curves, is the derivative, and the central concept of integral calculus, which originated in the calculation of areas and volumes, is the integral. Differential and integral calculus are connected by the fact that anti-differentiation, the operation inverse to differentiation, can be used to compute integrals.

Calculus, which has numerous applications in mathematics as well as other subjects, developed enormously in the eighteenth century, which is sometimes known as the "Century of Analysis." The two dominant figures in this development were the mathematician Joseph-Louis Lagrange (1736–1813) of Italian origin and, particularly, the Swiss mathematician Leonhard Euler (1707–1783). Euler approached calculus from a geometric viewpoint, advanced the concept of a function, defined the trigonometric functions (such as sine and cosine) without using geometry, introduced complex exponentials, and proved identities such as $e^{ix} = \cos x + i \sin x$, in addition to numerous other mathematical achievements. Euler and others also calculated expansions of functions into power series (for example, $\cos x = 1 - x^2/2! + x^4/4! - x^6/6! + \cdots$) and sums of infinite series (for example, $1 - 1/2^4 + 1/3^4 + 1/4^4 + \cdots = \pi^4/90$). They also studied differential equations (which are equations relating an unknown function $f(x)$ to its derivatives $f'(x), f''(x), f'''(x), \ldots$; a simple example is the equation $f'(x) - 2f(x) = 0$, whose solutions are functions of the form

Girolamo Cardano

$f(x) = ke^{2x}$) and partial differential equations (which are differential equations in which the unknown function depends on several variables).

Isaac Newton

The Nineteenth Century: the Emergence of Rigor and Abstraction

The development of mathematics accelerated considerably in the nineteenth century, and this acceleration affected every branch of mathematics. An immensely important figure of this era—and of all time—is Carl Friedrich Gauss (1777-1855), who is sometimes known as "The Prince of Mathematics." His contributions involve almost every branch of mathematics. As just one of his many achievements, Gauss gave in his 1799 thesis the first correct proof of the fundamental theorem of algebra, which states that any polynomial $a_0 + a_1x + a_2x^2 + \cdots + a_nx^n$ with real coefficients has at least one complex root (in other words, there is a complex number x that makes the expression equal to zero). Around 1825, the Norwegian Niels Henrik Abel (1802-1829) proved that it is impossible to solve a general polynomial equation with degree five or higher using just the five operations of addition, subtraction, multiplication, division, and extraction of roots. Other notable advances in algebra during this era were the work by Évariste Galois (1811-1832) on algebraic equations, which introduced the fundamental algebraic topics of group theory and finite fields; the invention by Richard Dedekind (1831-1916) of abstract structures called ideals; and the development of linear algebra, the study of objects such as vector spaces and matrices, by Arthur Cayley (1821-1895) and Leopold Kronecker (1823-1891). The evolution of algebra during this century illustrates the general movement of mathematics towards an abstract field detached from applications and physical intuition. During this era, mathematicians were not afraid to invent abstract objects and to study these objects for their own sake.

Another development of nineteenth century mathematics was the drive for rigor. This was especially important in analysis, where mathematicians had been manipulating infinitesimally small numbers for more than a century without knowing if these manipulations were mathematically legitimate. One of the main achievements of Augustin-Louis Cauchy (1789-1857) was to establish the foundations of a rigorous framework for analysis. For example, Cauchy put the concept of limits at the basis of differential and integral calculus and formulated quite precise definitions of continuity and integration. In addition, Cauchy proved general theorems about the existence and uniqueness of

Leonhard Euler

Carl Friedrich Gauss

solutions to differential equations. However, rigor was not fully established in analysis until Karl Weierstrass (1815–1897) formulated his epsilon-delta definitions of continuity and other basic concepts (which every student today must have at the tip of his fingers) and rigorously constructed the set **R** of real numbers, which he completed around 1870.

Two areas of analysis, complex analysis and Fourier analysis, emerged during the nineteenth century and have important theoretical and practical applications. Complex analysis deals with functions of one or more complex variables. Fourier series deals with writing periodic functions $f(x)$ as an infinite sum of trigonometric functions of the form

$a_0 + a_1 \cos x + b_1 \sin x + \ a_2 \cos 2x + b_2 \sin 2x + a_3 \cos 3x + b_3 \sin 3x + \ldots$

In fact, it was certain problems about Fourier series that led Georg Cantor (1845–1918) to create set theory around 1880. This was another significant development of this era. In some sense, Cantor's set theory is the arithmetic of infinite numbers, as it provides a way to compare sizes of infinite sets. For example, this theory allows countably infinite sets (such as the set of integers) to be distinguished from uncountably infinite sets (such as the set of real numbers). This theory would become the foundation of all mathematics in the following century. Yet near the turn of the twentieth century, mathematicians realized that set theoretical concepts can lead to paradoxes, for example when one tries to determine the size of the set of all sets. In the following years, these paradoxes, along with the birth of mathematical logic, gave rise to very active research in the foundations of mathematics and the field of metamathematics.

Innovations proliferated in geometry as well. Jean-Victor Poncelet helped develop projective geometry, the study of geometric properties that are invariant under projection. Gauss, Nikolai Lobachevksy (1792–1856), János Bolyai (1788–1867), and Bernhard Riemann (1826–1866) gave birth to differential geometry and non-Euclidean geometry, which were freed from the restriction of Euclid's fifth postulate (which states that, given a line and a point not on the line, there is a unique line passing through the point and parallel to the given line) and not limited to the plane or three-dimensional space.

Arthur Cayley

Poincaré and Hilbert: Beacons at the Turn of the Century

Augustin-Louis Cauchy

Our flight through the evolution of mathematics has now brought us to the turn of the twentieth century. During these years, two geniuses dominated mathematics: the Frenchman Henri Poincaré (1854–1912) and the German David Hilbert (1862–1943). It is said that these were the last universal mathematicians, in the sense that they were active or at least well-informed in almost every area of mathematics. Poincaré greatly contributed to classical analysis (the theory of elliptic functions, differential equations, etc.), developed quantitative tools for exploring the behavior of systems of differential equations (which anticipated modern chaos theory), contributed to the development of topology and algebraic topology, and worked on celestial mechanics and relativity. Hilbert's work dealt with the theory of invariants, the theory of algebraic numbers (which are roots of a polynomial with integer coefficients), the axiomatization of geometry, infinite dimensional vector spaces, metamathematics, and problems in mathematical physics, among other subjects.

The importance of Poincaré and Hilbert reflects the importance of the French and German schools of mathematics, which were the most renowned in Europe at the time. The early twentieth century was a golden age of French mathematics; in addition to Henri Poincaré, French mathematics benefited from the brilliance of mathematicians such as Émile Picard (1856–1941), Jacques Hadamard (1865–1963), Émile Borel (1871–1956), René Baire (1874–1932), and Henri Lebesgue (1875–1941), who created a more general and powerful theory of integration than the one developed by Cauchy and Riemann. Almost all French mathematicians of this era were analysts whose research concentrated on function theory. One notable exception is Élie Cartan (1869–1951), who worked at the crossroads of group theory, systems of differential equations, and geometry. Cartan's work, which was ahead of its time, remained unrecognized for many years.

Poincaré, who benefited from strong geometric intuition while being criticized for lacking clarity and rigor, usually worked alone; and because he had no students, Poincaré failed to found a school of mathematics. Meanwhile, Hilbert had a completely different style. He taught many students and inspired his fellow mathematicians. Hilbert's work (which was more algebraic and rigorous than Poincaré's) was remarkable and his perspective on mathematics was insightful. For example, in a lecture at the International Congress of Mathematicians in Paris in 1900, Hilbert proposed his famous list of twenty-three important

Georg Cantor

Henri Poincaré

problems to be solved, which inspired much research throughout the following decades. Above all, Hilbert turned the University of Göttingen into the world capital of mathematics for the first three decades of the twentieth century. He was able to achieve this in part because of the mathematical talent he attracted to the university, but mostly because of the stimulating and friendly atmosphere he inspired there. This prestigious center of mathematics developed a powerful and modern school of abstract algebra, whose key figures included Émil Artin (1898-1962), Emmy Noether (1882-1935)—the daughter of the mathematician Max Noether—and the Dutch mathematician Bartel L. van der Waerden (1903-1996).

While the reign of Göttingen lasted until 1933—when it was suddenly faced with the Nazis and their antisemitism—the golden age of mathematics in France ended much earlier. World War I and the postwar years marked the decline of mathematics as well as of other scientific fields, including physics.

What was the scale of this decline? What caused it? Even the historians of today haven't fully answered these questions, but it is certain that this decline is what caused Bourbaki to form. The principal reason for the decline—and the reason that Bourbaki itself referred to most—seems to be the loss of scientists in World War I. In his *Souvenirs d'apprentissage*, André Weil wrote that "already at l'École [Normale] I was struck by the damage that World War I had done to mathematics. The war had created a void, and it was not easy for my generation and the younger ones to fill it. In 1914, the Germans wisely tried to save their future scientific elite, and they managed to protect many young scientists. In France, a movement for equality in the face of sacrifice, although certainly praiseworthy in principal, led to completely opposite policies. The monument to the dead at l'École Normale is just one illustration of the disastrous consequences." The numbers collected by the mathematician Martin Adler of the University of Versailles are powerful: almost half of the mathematics students who entered l'École Normale Supérieure between 1911 and 1914 died in the war. Overall, a fourth of the 331 students who entered the ENS between 1900 and 1918 did not return from the battlefield. It is likely that other educational institutions suffered similar losses. Jean Dieudonné, a member of Bourbaki, described in an interview with Marian Schmidt how "[...] it was the young mathematicians killed in the war who would have been able to continue the work of Poincaré and Picard. My generation suffered the consequences of this interruption, which lasted around fifteen years. Our professors were all twenty or thirty years older than us and taught us mathematics from their youth rather than new theories. Ideally, a math professor should

David Hilbert

be not more than ten years older than his students, or at the very most fifteen years older, so that he can teach the mathematics of his students' era. Without meaning to boast, I can say that it was Bourbaki that saved French mathematics from extinction."

Apprentices without Masters

Yet the problems of French mathematics can not be attributed entirely to the wartime loss of young scientists, since Germany certainly lost young talent to some extent. As other factors in the decline of French mathematics, Liliane Beaulieu suggests the rigidity of French scientific institutions, which was greater than that of their less centralized German counterparts, and insufficient funding after the war, which created a competitive hierarchy in French science. In 1938, for example, Weil complained bitterly about the "high priests" of French science, who used their power to monopolize collaborators and to distribute funds while being "strangers to the real problems, to the living ideas of modern science."

When budding mathematicians entered l'École Normale Supérieure in the 1920s, they were met by an antiquated faculty focused

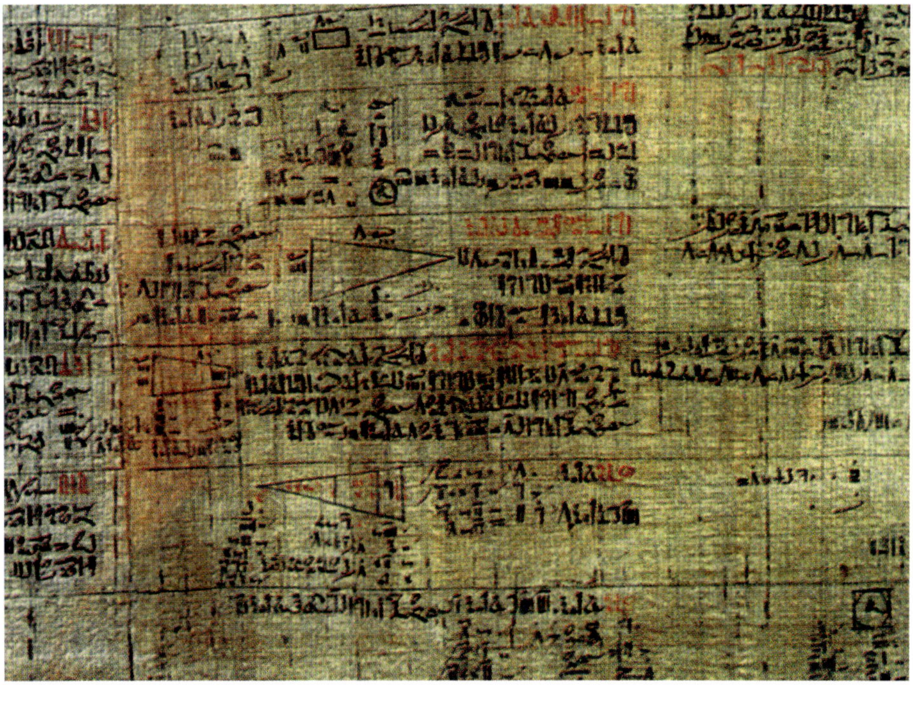

The Rhind Papyrus, held in the collections of the British Museum, dates from around 1650 B.C. It contains calculations with fractions and solutions to various arithmetic and geometric problems.

Emmy Noether

Christophe (an author who had studied at l'École Normale) used the famously absent-minded mathematician Hadamard as a model for the character Professor Cosinus, who was fond of "imaginary circles with indeterminate centers." The caption states that "having borrowed an almost new coat, he decides to return home, although not before promising his gracious hostess to dedicate to her his next work on the geometric location of the intersection of imaginary circles with indeterminate centers."

primarily on "Daddy's function theory," as the founders of Bourbaki later described it. As André Weil wrote in the *Gazette des mathématiciens* in 1991, these apprentice mathematicians "were left essentially without masters." They worked mostly among one another, each giving lectures for the other students. According to Weil, "We learned much more from each other than from the classes we went to—or skipped." Poincaré died relatively early, in 1912, without leaving any intellectual descendants. Meanwhile, Élie Cartan, whom the future members of Bourbaki respected (independently of the fact that his son Henri was a member), was difficult to understand and quite isolated in his work. Almost the only mathematician of the older generation to convey some fresh ideas in his lectures was Hadamard, who organized a seminar for the students. While being an analyst, Hadamard was very open and benefited from his broad scientific education. "It was Hadamard who made me into a mathematician. He was full of ideas and interested in everything, even in number theory, which wasn't taught at that time [in France]," Weil writes.

Finding postwar France to be nearly devoid of living mathematics, some young Normaliens went to study elsewhere. Weil was one of the first, if not the very first, to do so, and he set an example for others to follow. In 1925, when he was only nineteen years old, Weil went to Italy and then to Germany for one year on a fellowship from the Rockefeller Foundation. It was only Weil's great independence that allowed him to go on such a journey, since nationalist hatred prevented normal scientific relations with Germany and its allies. In fact, it was not until 1928 that a delegation of German scientists, led by Hilbert, was allowed to attend the International Congress of Mathematicians, held that year in Bologna.

In addition to Weil, seven other future participants in Bourbaki's early years completed their studies abroad. Szolem Mandelbrojt went to Rome from 1924 to 1925 and then the United States for one year; Paul Dubreil lived in Germany and then Italy from 1929 to 1931; Claude Chevalley studied in Hamburg from 1931 to 1932 and in Marburg during the summer of 1933; René de Possel spent a few months in Munich in 1930–1931, followed by a month in Hungary before returning to Germany to study in Göttingen and Berlin; Jean Dieudonné went to Princeton for the 1928–1929 academic year and then to Berlin and Zurich; Jean Leray travelled to Berlin in 1932 and then to Leipzig, but was forced to cut his trip short (he was planning to go to Göttingen) due to the political situation. In general, these trips were extremely beneficial, since they let the vitality of the German algebraists inspire at least some of the future members of Bourbaki. Indeed, the German school of algebra had a strong influence on Bourbaki, whose treatise in some sense "algebrized" mathematics. Later, the members of Bourbaki often asserted that they were inspired by Hilbert, more so than by Poincaré, whose genius they recognized but whose style they disliked.

Charles Ehresmann

Down with *Cours d'Analyse!* Long Live *Moderne Algebra!*

The influence of German algebraists on Bourbaki was partly thanks to the book *Moderne Algebra*, written in German by the Dutch mathematician van der Waerden and published in 1930–1931. Van der Waerden was one of Emmy Noether's disciples at Göttingen, and he based his book partly on lectures by Emmy Noether and Emil Artin, another eminent German algebraist. *Moderne Algebra* was an immediate success. It impressed the future founders of Bourbaki, not just because of the modern subjects it discussed but also because of its concise and rigorous presentation that emphasized general structures like groups, rings, and fields. As Jean Dieudonné said in 1968, van der Waerden's book "is characterized by extremely precise language. The organization of the ideas and different parts of the book is extremely tight."

In terms of style, this book contrasted sharply with French textbooks like Édouard Goursat's *Cours d'analyse*. In his book, Goursat discussed classical analysis in great detail, even in too much detail, since it resulted in a verbose text that concealed the general concepts. The book lacked rigor, and theorems were repeated, each time with additional hypotheses. Furthermore, Goursat completely ignored recent developments in analysis such as Lebesgue's theory of integration, which dated from around 1902.

In deciding to write a modern treatise on analysis, the Bourbaki group had given itself the task of "cleaning the stables of Augias" (as Weil described it, referring to the fifth labor of Hercules). They had to refine the presentation, strengthen the rigor, and include more recent developments. But the treatise achieved much more than the group had set out to do.

In 1934, when the idea of forming the group was born, the group's future members were already experienced mathematicians. Too young to hold professorships in Paris, most of the future members enjoyed the freedom that went with their professorships at universities outside the capital. These were not mathematicians of marginal talent, and the older generation recognized their scientific merits and honored them with various distinctions such as awards from the Académie des Sciences. At the same time, these young mathematicians did not hold

Jacques Hadamard (1865–1963)

great respect for their professors, who were more than a generation older than them. Most of them benefited from studying abroad, where they met notable mathematicians and encountered new perspectives on mathematics. By learning how mathematics was more alive abroad than in France, these young mathematicians became increasingly aware of the antiqueness of "Daddy's function theory" and of the deterioration of French science and its "high priests." The plan to write a treatise on analysis was a plan to revive French mathematics, although they were not completely aware of this at the time. Their talents, the diversity of their experiences and specialties, the influence of German algebraists, and their fierce desire to clarify necessary concepts all contributed to the fact that the Bourbakis achieved much more than simply writing a treatise on analysis. The group not only succeeded in reforming analysis but in reforming mathematics as a whole. However, at the time the young Bourbakis had not yet realized the extent of their future influence.

B. L. Van der Waerden's *Moderne Algebra* (1930–1931) served as a model for Bourbaki.

In front of the Chanteloup Pagoda near Tours (from left to right): Jean Delsarte, René de Possel, and Szolem Mandelbrojt.

The "Medal Wars"

The young Bourbakis didn't have a high opinion of the French scientific institution and how it functioned, and in particular Weil and Delsarte expressed their opinions on this subject in the late thirties. The episode of the "Medal Wars" provides a good illustration of how these opinions played out.

The affair started with Jean Perrin, a physicist and Nobel Prize winner, who had been the Undersecretary of State for Scientific Research since 1936. Perrin, the future creator of the CNRS (Centre National de la Recherche Scientifique, the French scientific research institute), was a "powerful and fiery" character who was influential in a certain "group of members of the Institute [who] held most of the control levers of French scientific life," as Weil describes in his biographical note on Delsarte. In 1937, Perrin devised "a system of rewards, both honorary and monetary, graded from modest little medals to the great Médaille d'Or worth 250,000 francs." The rewards totalled over two million francs, an enormous amount of money for that time. The decree, signed by the Minister of National Education Jean Zay, was published in 1937 during Bourbaki's summer conference in Chançay. According to Weil, Delsarte realized that this would create "more pettiness and scheming in a system already far from being free of these." More precisely, Liliane Beaulieu explains, those who opposed this plan believed that the large sums involved in these medals would accentuate the disparities between scientists (by making the rich richer), give more power to Perrin and his clique, and increase quarrels among scientists.

At the beginning of the academic year, Delsarte, Weil, and some others took advantage of the fact that the decree could not be implemented until funds were allocated for the prizes and distributed petitions against the medal system throughout the French universities. Delsarte, Weil, and two other academics brought the petitions, which bore almost 400 signatures, to the Minister Jean Zay in person. But these efforts were in vain, and, apparently, it even led to threats against certain people who signed the petitions. Nevertheless, in the end Delsarte and his friends won the fight. During the Parliament's vote on the budget on the night of December 31, "Jean Perrin defended his plan in the House until the small hours of the morning, while in the Senate the famous Joseph Cailloux, always looking for ways to economize, removed the two million francs for the medals every time he saw them reappear." After some confusion, it turned out two days later that the budget for the medals had been rejected.

Jean Perrin

But this victory was fleeting. The CNRS has had its system of medals since the fifties—and, apparently, without causing any real problems. Could the scientists of the late twentieth century be more corrupt than those of the 1930s? Or could the harmful effects of the medal system have been allayed by the less elitist system of today? These medal wars contrasted two concepts on how to finance research: one rewarding scientists for their results, which would help already well-established researchers, and the other encouraging young students to pursue a career in research, which would rely on betting on the future. The creation of the CNRS, which institutionalized the research profession, defused the opposition against rewards.

4

Bourbaki's Éléments de Mathématique

Bourbaki's work is a monumental treatise of more than seven thousand pages. The first volume was published in 1939 and the most recent in 1998. Does the future hold more? Many doubt it.

Admired by some, hated by some, Bourbaki's treatise *Élements de mathématique* (published in English as *Elements of Mathematics*) met success and influenced the mathematics of its time. Recall that the mathematicians who wrote this treatise initially aimed to write a modern textbook on mathematical analysis that would cure the problems of mathematical education in French universities and of the current analysis textbooks. In the first month of the group's existence, even before it had been officially formed, the Bourbaki group immersed itself in profound thought and discussion about the content of their future book.

The volumes of *Éléments de mathématique* (most recent editions) from the library at the Henri Poincaré Institute.

What topics did the young Bourbakis plan to include in their book? Of course, they planned to discuss classical topics of analysis such as analytic functions, Fourier series, differential equations, and integration, almost all of which were included in the existing textbooks. But they also wanted to modernize the discussion of these topics. To achieve this, the Bourbakis had decided during their preliminary meetings of 1934 and 1935 to include some general concepts that would be useful throughout their books. In particular, they decided to include some essential concepts from set theory and modern algebra, following van der Waerden's *Moderne Algebra*, which had been published in Germany in 1930–1931 and was greatly appreciated by the future Bourbakis and other mathematicians. This "abstract packet," as they called the set of these fundamental tools and concepts, was to include only the absolutely necessary topics. At this time, the Bourbakis did not plan to include abstract topics in their own right, and they did not attach any special mathematical vision or ideology to the project.

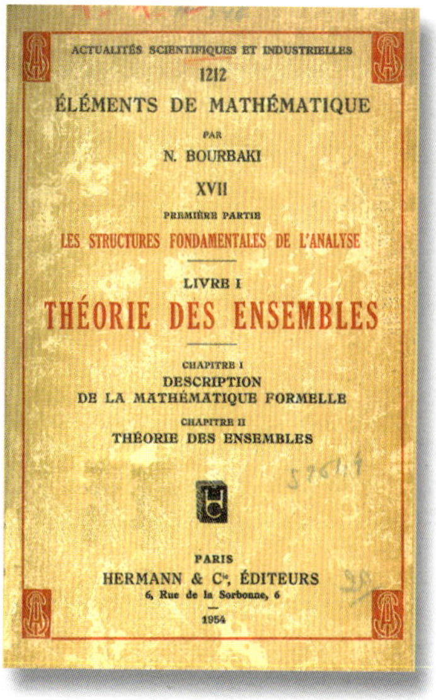

The first volume of the book on set theory.

The young mathematicians added more and more topics as they developed a detailed outline of the treatise. Already in 1935, René de Possel suggested that *Traité de mathématiques* would be a more appropriate title than *Traité d'analyse.* At the founding conference in Besse-en-Chandesse in July 1935, the Bourbakis decided to present their material axiomatically—that is, they would precisely and clearly state the set of basic rules (axioms) that the mathematical objects in question must obey and then explore the properties (theorems) following logically from these rules. However, at this point, the group still planned to limit the preliminary content as much as possible.

Since the Bourbakis wanted the abstract packet to contain the concepts used for the main topics of the treatise, they started their work on the treatise by writing the abstract packet. The abstract packet grew endlessly over the course of the long and tortuous process of writing this preliminary section. Meanwhile, the Bourbakis were developing a new vision of mathematics that saw this discipline as an edifice endowed with profound unity, founded on set theory and built in terms of abstract structures.

Bourbaki's initial project thus found itself expanded in terms of both length and spirit. By 1941, the outline for the treatise was divided into four parts, with each part divided into books and each book divided into chapters. These four parts were called Fundamental Structures of Analysis (eight books), Functional Analysis (seven books), Differential Topology (two books), and Algebraic Analysis (eight books). The books planned for the first part, for example, were I. Set Theory, II. Algebra, III. General Topology, IV. Topological Vector Spaces, V. Differential Calculus, VI. Integration, VII. Combinatorial Topology and Differential Varieties, and VIII. Analytic Functions. Considering that the first part alone would contain copious material and the heart of the work would be formed by the other three parts, it is clear that the group's simple treatise on analysis had become a massive project.

Ten Books and Over Sixty Chapters

The group's ambitious plans have only partially materialized. Over sixty-five years after the birth of Bourbaki, today *Éléments de mathématique* consists of ten books (with most books initially published in several volumes). The group's ten books, listed with the French title followed by the title of the English translation where one exists, are

- *Théorie des ensembles* (*Theory of Sets*, a summary of results plus four chapters)

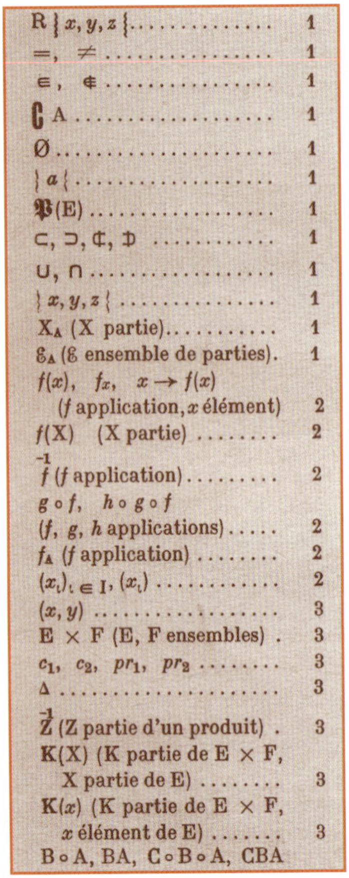

$\mathrm{R}\{x, y, z\}$	1
$=$, $\neq$	1
$\in$, $\notin$	1
$\complement \mathrm{A}$	1
$\emptyset$	1
$\{a\}$	1
$\mathfrak{P}(\mathrm{E})$	1
$\subset$, $\supset$, $\not\subset$, $\not\supset$	1
$\cup$, $\cap$	1
$\{x, y, z\}$	1
X_{A} (X partie)	1
$\mathfrak{E}_{\mathrm{A}}$ ($\mathfrak{E}$ ensemble de parties)	1
$f(x)$, f_x, $x \rightarrow f(x)$ (f application, x élément)	2
$f(\mathrm{X})$ (X partie)	2
$\overset{-1}{f}$ (f application)	2
$g \circ f$, $h \circ g \circ f$ (f, g, h applications)	2
f_{A} (f application)	2
$(x_\iota)_{\iota \in \mathrm{I}}$, (x_ι)	2
(x, y)	3
$\mathrm{E} \times \mathrm{F}$ (E, F ensembles)	3
c_1, c_2, pr_1, pr_2	3
Δ	3
$\overset{-1}{\mathrm{Z}}$ (Z partie d'un produit)	3
$\mathrm{K}(\mathrm{X})$ (K partie de $\mathrm{E} \times \mathrm{F}$, X partie de E)	3
$\mathrm{K}(x)$ (K partie de $\mathrm{E} \times \mathrm{F}$, x élément de E)	3
$\mathrm{B} \circ \mathrm{A}$, BA, $\mathrm{C} \circ \mathrm{B} \circ \mathrm{A}$, CBA	

An extract from the index of set theoretical notation, some of which was invented by Bourbaki and adopted by the mathematical community.

Jacques Herbrand (1908–1931)

- *Algèbre* (*Algebra*, ten chapters)
- *Topologie générale* (*General Topology*, ten chapters)
- *Fonctions d'une variable réelle* (*Functions of One Real Variable*, seven chapters)
- *Espaces vectoriels topologiques* (*Topological Vector Spaces*, five chapters)
- *Intégration* (*Integration*, nine chapters)
- *Algèbre commutative* (*Commutative Algebra*, ten chapters)
- *Variétés différentielles et analytiques* (a summary of results on differential and analytic varieties with proofs omitted)
- *Groupes et algèbres de Lie* (*Lie Groups and Lie Algebras*, nine chapters)
- *Théories spectrales* (a book of two chapters about spectral theory)

See pages 60–68 for an introduction to the branchs of mathematics discussed in these ten books.

Initially, the first six of the books listed above formed the first part of *Éléments de mathématique*, but the plan to divide the work into parts was abandoned when the Bourbakis realized after a few years that the three other parts would long remain incomplete. All the books published before 1958 belonged to this first part.

The first volume to appear was the summary of results in set theory, published in 1939. Since Bourbaki believed that set theory was the foundation for all of mathematics, it makes sense that this is the subject they dealt with first. But their task proved to be arduous, and the group was impatient to publish. Because of this, the Bourbakis decided to begin with publishing a summary of results without their proofs, a book which would certainly be useful.

This collection of basic results was published by Hermann, a publishing house in Paris. At the time, the publishing house Gauthier-Villars monopolized French mathematical publication, so why did Bourbaki choose Hermann? The answer, as André Weil explains in his *Souvenirs d'apprentissage*, is that "we weren't at all tempted to turn to [Gauthier-Villars]; it was far too academic for us." In fact, among the directors of Gauthier-Villars were scientists such as Émile Picard and Émile Borel, who were well-established mathematicians belonging to the very generation that Bourbaki expected so little from and who surely looked unfavorably upon the projects of our young Turks. Instead of these, in 1935 Bourbaki chose the future editor of *Éléments de mathématique* as Weil's friend Enrique Freymann, the director of the Parisian publishing house Hermann.

Freymann, Editor in Crime

Of Mexican origin, Freymann was a colorful character who had taken over the business of his father-in-law Hermann upon his death. According to Claude Chevalley's remarks collected by Judith Friedman, Freymann was a charming person who loved talking with students and professors who stopped by his store, and Weil writes that "I hardly every went there without a few hours at my disposal, and I always left reluctantly." Freymann had worked with Weil and Chevalley when these two decided to publish a collection of mathematical articles in memory of their friend Jacques Herbrand, a brilliant logician who died tragically in a mountaineering accident in 1931 at the age of just 23. Hermann published these articles, which were written by various mathematicians including Emmy Noether, Helmut Hasse, and John von Neumann, in 1934–1935. They appeared in the series *Actualités scientifiques et industrielles* ("Current events in science and industry"), which Weil described as "a sort of spider web that [Freymann] used to attract both the elite and the underworld of international science into his lair. He was open to all ideas, from the most serious to the most preposterous."

Freymann played a crucial role in the launching of Bourbaki. According to Chevalley, it was Freymann who united Weil and himself with the idea of writing a treatise on analysis to replace Goursat's obsolete book. Whether or not this was the case, Freymann did wholeheartedly support Bourbaki's enterprise despite the financial risk that it carried. As Weil writes in his autobiography, "As soon as we mentioned the idea of Bourbaki, he didn't hesitate for a second [...]. And he didn't have reason to regret his decision to trust us and to encourage us constantly from the very beginning, since Bourbaki did become one of Hermann's major financial assets. Yet his decision to join our adventure was no small virtue [...], as there were plenty of well-intentioned pedants who warned him that it was just another one of the Normaliens' pranks and that falling for it would make him look ridiculous. In fact, perhaps the name and the legend of Nicolas Bourbaki, which he zealously helped develop and spread, were not the least of our project's attractions for him."

The group expressed its profound gratitude for Freymann in the dedication of the first volume of *Théorie des ensembles* (*Theory of Sets*), published in 1954, as Freymann had died during the proofreading of this book. Such a dedication is no small matter when one considers the impersonal character of Bourbaki's treatise.

Hermann and Bourbaki's honeymoon lasted until the early 1970s, seeing the publication of over thirty volumes of *Éléments de*

Extrait de LA TRIBU N° 29
(Congres de l'incarnation de l'Ane qui trotte)
Celles-sur-Plaine (1952)
Air de «En descendant la rue d'Alger»

Le mod' d'emploi de ce Traité (bis)
Est d'la plus grand' simplicité (bis)
Si quelqu'chos' n'est pas clair
Eh bien
Il vous suffit d'abstraire
Et la lumière revient (bis).

L'alphabet de tout' les nations (bis)
Sera mis à contribution (bis)
On a, pour être plus clair
Souvent
Mis en p'tits caractères
Les passages importants (bis).

Les exercic' sont alléchants (bis)
Leur énoncé est séduisant (bis)
N'essayez pas d'en faire
Beaucoup
Car on dit qu'les deux tiers
Sont faux ou au d'ssous d'tout (bis)

Les notations, vous le verrez (bis)
Ont été fort améliorées (bis)
Et le meilleur critère
Eh bien
C'est que sur tout'la Terre
Personn' n'y comprend rien (bis).

L'ordre suivi dans l'exposé (bis)
A été longu'ment médité (bis)
On met les points s'condaires
Devant
Et puis, en corollaires
C'qui doit servir tout l'temps (bis).

The lyrics of the original song *En descendant la rue d'Alger* can be found at http://homepages.ulb.ac.be/~xhubaut/chorale/belepoq.htm
(For English translation, see appendix.)

VI MODE D'EMPLOI DE CE TRAITÉ

dérations n'apparaîtra donc au lecteur que s'il possède déjà des connaissances assez étendues, ou bien s'il a la patience de suspendre son jugement jusqu'à ce qu'il ait eu l'occasion de s'en convaincre.

4. Pour obvier en quelque mesure à cet inconvénient, on a inséré assez souvent, dans le cours du texte, des exemples qui se réfèrent à des faits que le lecteur peut déjà connaître par ailleurs, mais qui n'ont pas encore été exposés dans le traité ; ces exemples sont toujours placés entre deux astérisques *.... *. La plupart des lecteurs trouveront sans doute que ces exemples leur faciliteront l'intelligence du texte, et préféreront ne pas les omettre, même en première lecture ; une telle omission, néanmoins, n'aurait bien entendu, du point de vue logique, aucun inconvénient.

5. Le lecteur voudra peut-être parfois se faire une idée sommaire de l'ensemble d'un Livre, avant d'en aborder l'étude détaillée. Cette tâche lui sera facilitée par des *fascicules de résultats*, annexés, en principe, à chaque Livre, et destinés également aux lecteurs pressés, qui désireraient arriver le plus vite possible à l'étude de problèmes spéciaux. Ces fascicules contiendront, autant que possible, l'essentiel de ce qui sera nécessaire à l'étude des parties suivantes.

6. L'armature logique de chaque chapitre est constituée par les *définitions*, les *axiomes* et les *théorèmes* de ce chapitre : c'est là ce qu'il est principalement nécessaire de retenir en vue de ce qui doit suivre. Les résultats moins importants, ou qui peuvent être facilement retrouvés à partir des théorèmes, figurent sous le nom de « propositions », « lemmes », « corollaires », « remarques », etc. ; ceux qui peuvent être omis en première lecture sont imprimés en petits caractères. Sous le nom de « scholie », on trouvera quelquefois un commentaire d'un théorème particulièrement important.

7. Certains passages sont destinés à prémunir le lecteur contre des erreurs graves, où il risquerait de tomber ; ces passages sont signalés en marge par le signe Z (« tournant dangereux »).

An extract for the instructions for using Bourbaki's treatise, including an explanation of the rounded Z symbol.
(For English translation, see appendix.)

mathématique. Unfortunately, their relations were subsequently poisoned by a long dispute over royalties, as royalty payments were delayed and translation rights were granted without Bourbaki's knowledge. The dispute reached the courtrooms, and the verdict given in 1979 hit the publishers hard: Bourbaki recovered the material and intellectual property rights for all of its books. However, the conflict caused a long interruption in the publication of Bourbaki's books. When publication did resume in 1980, it was not by Hermann but by Masson, which now distributes all of Bourbaki's books published to date.

Now let's return to the treatise itself. What is so special about *Éléments de mathématique*? The three-page text entitled "Instructions for Using This Book" and printed at the beginning of every volume reveals some of this treatise's unique characteristics.

According to the first line of these instructions, "This treatise starts with the basis of mathematics and provides complete proofs." This announces the book's style: the treatise starts from the very beginning and will prove every theorem. This might make a beginner think, "Terrific! I'll learn all of mathematics systematically from A to Z without skipping anything!" But one who thinks so should think again. In fact, as the instructions continue, "In principal, this book does not assume any particular mathematical knowledge. It does, however, assume a certain level of mathematical maturity and reasoning ability." This is a euphemism. According to its instructions, the treatise is intended for anyone having properly assimilated his first- or second- year university classes. This underestimates the necessary prerequisites unless the reader is extremely talented, motivated, and inclined towards abstraction. The truth is that in general only professors and graduate students can benefit from Bourbaki's books. It's far from a work for the general public!

However, neither is it a research work in which Bourbaki discusses new results proven by the group. In fact, Bourbaki did not really produce new mathematical discoveries or inventions as a group, although the Bourbakis published new results as individuals and the group's treatise contains, according to the mathematician Jean-Pierre Bourguignon, "some superb and very clever proofs." With just a few exceptions, *Éléments de mathématique* is in fact a broad summary of mathematics, a modern reorganization and reformulation of existing knowledge. Indeed, the word *Éléments* used in the title echos Euclid's famous *Elements*, just as the style of Bourbaki's treatise echos that of Euclid's. Also, the use of the unusual singular *mathématique* in place of the usual plural *mathématiques* emphasizes Bourbaki's belief in the profound unity of mathematics.

The treatise should be read in a certain order. Thus in the first six books (*Théorie des ensembles*, *Algèbre*, *Topologie générale*, *Fonctions d'une variable réelle*, *Espaces vectoriels topologiques*, and *Intégration*), "each statement depends only on definitions and results stated earlier in the book or in previous books." On the other hand, the later books (*Algèbre commutative*, *Variétés différentielles et analytiques*, *Groupes et algèbres de Lie*, and *Théories spectrales*) assume the knowledge of the first six books but do not follow any order among themselves. One should note that the books were not published in their logical order. For example, the first two chapters of the book on set theory, which formed the very beginning of the treatise's logical progression but created difficulties for the Bourbakis, did not appear until 1954, long after the books on algebra and topology had been published.

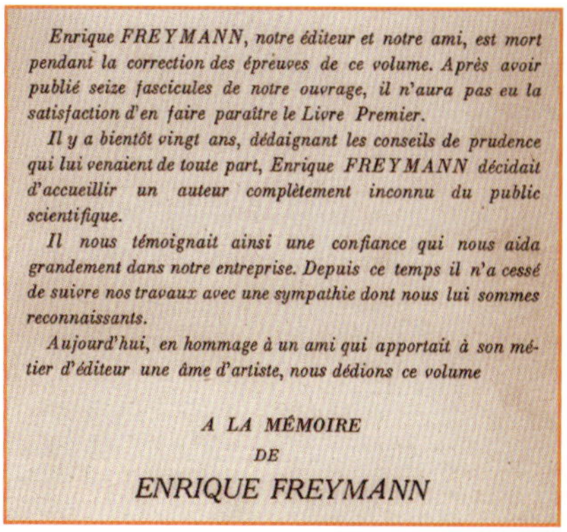

Enrique FREYMANN, notre éditeur et notre ami, est mort pendant la correction des épreuves de ce volume. Après avoir publié seize fascicules de notre ouvrage, il n'aura pas eu la satisfaction d'en faire paraître le Livre Premier.

Il y a bientôt vingt ans, dédaignant les conseils de prudence qui lui venaient de toute part, Enrique FREYMANN décidait d'accueillir un auteur complètement inconnu du public scientifique.

Il nous témoignait ainsi une confiance qui nous aida grandement dans notre entreprise. Depuis ce temps il n'a cessé de suivre nos travaux avec une sympathie dont nous lui sommes reconnaissants.

Aujourd'hui, en hommage à un ami qui apportait à son métier d'éditeur une âme d'artiste, nous dédions ce volume

A LA MÉMOIRE
DE
ENRIQUE FREYMANN

The dedication to Enrique Freymann in the first volume of the book on set theory, published in 1954.
(For English translation, see appendix.)

From General to Specific

One important characteristic of the treatise is that "the presentation is axiomatic and usually progresses from general to specific." The presentation could have remained axiomatic even if each topic were introduced with a few concrete examples motivating the abstract concepts. However, Bourbaki rejected this approach in favor of a purified presentation in which sparse examples followed the abstract concepts. This was not a brilliant decision from a pedagogical point of view. So, while Bourbaki warned in the instructions that "the reader may not understand the use of some of the material until later chapters, unless he already has a reasonably broad knowledge of mathematics," many complained about this lack of motivation. An author could find such a style more simple and concise to write; the reader, on the other hand, suffers no matter how clear the presentation.

To try to solve this problem to some extent, Bourbaki included many notes to place the topics in their historical context. These historical notes, usually found at the end of a chapter, owe their existence to Weil and Dieudonné, who were greatly interested in the history of

Helmut Hasse (1898–1972).

mathematics. These notes were later collected and published as a single volume called *Éléments d'histoire des mathématiques* (published in English as *Elements of the History of Mathematics*). Note that *Éléments d'histoire des mathématiques* uses the plural *mathématiques,* whereas *Élements de mathématique* uses the singular, implying that Bourbaki unified mathematics from multiple subjects into a single one.

Each of the treatise's chapters also contained exercises. Dieudonné contributed so much to these exercises that Bourbaki honored him semi-officially in the preface to the second edition of *Topologie générale* by writing that "as always, I am also indebted to my faithful adjudant for most of the exercices." The solutions to several of the problems were lost when Dieudonné died, and it would not be an easy task to reconstruct these answers. Bourbaki's exercises, whose quality is often praised, serve two purposes. One purpose is, as usual, to let the reader check if he understands the material. The other is to "show him results not found within the body of the text." In other words, an interesting (and important) result was often exiled to the exercise section if it was an application rather than a step in the strictly logical development of the text. This is one of the weaknesses of the treatise. Another weakness is the bibliographical references, which Bourbaki included sparingly, "the text being devoted to the dogmatic presentation of a subject." The references are grouped together at the end of the historical notes and usually contain "only original books and papers that had the most important influence on the evolution of the topics in question." Many readers would have wished for more.

New Terminology, New Notation

Émil Artin (1898–1962) was a notable German algebraist.

One of the treatise's virtues is that the Bourbakis devoted copious attention to terminology, as their goal of using the most rigorous and simple language possible led them to develop numerous new terms. For example, they invented the word "bijection," which refers to a correspondance between two sets that associates each element of the first set to an element of the second set and vice versa. In addition to terminology, Bourbaki invented new notation. The most famous example is the symbol $\varnothing$, which represents the empty set (that is, the set with no elements). This symbol was invented by Weil, the only Bourbaki to be familiar with the Norwegian alphabet. Another famous symbol invented by Bourbaki is a sort of large, rounded Z (meaning "dangerous curves ahead") that the Bourbakis placed in the margin of their treatise whenever they wanted to warn the reader about a serious error made by many students. Although not every term and symbol earned

1. Dieudonné on How the Treatise Was Written

At a 1968 conference in Romania, Dieudonné described the process of writing *Éléments de mathématique* as follows:

"Although Bourbaki's method of working is terribly long and arduous, the goals of the work almost require such a method. Once the group basically agrees on the need to write a chapter on such and such a theory [...], at one of the conferences (which are held about three times a years) the chapter is assigned to a member who would like to work on it. He then writes a first draft of the chapter or chapters that were assigned to him, basing his work on a vague outline developed by the group. He can change the content of the outline more or less as he pleases, adding what he likes and deleting what he doesn't. However, he does this at his own risk, as you will soon see. After a year or two, the completed draft returns to one of the Bourbaki conferences, where it is usually read outloud, without skipping a single page. The group examines each proof step by step and mercilessly criticizes the work being read. It's impossible for someone who's never been to a Bourbaki conference to understand how vicious this criticism can be. It exceeds by far all the criticism outside people have made against Bourbaki, and it is often delivered in language that I couldn't possibly repeat here [...].

Once the group has torn the first draft apart into tiny pieces, really annihilated the work, it assigns it to a second member to start over. This poor soul knows what awaits him, since even if he follows the group's new instructions, the group's ideas will change while he is writing the new draft, and the draft will probably be torn apart once again the following year. At this point, a third member starts a third draft, and so on. One might think that this process will never end, that it is a recurrence with no reason to terminate. Indeed, we end the process for purely human reasons. After six or seven or eight or ten times of seeing the same chapter, everyone is so exhausted that the vote to publish the chapter is unanimous." This last sentence is most likely an exaggeration. It would probably be more accurate to say that if the *n*th version doesn't earn some minimum level of support, the chapter would remain shelved, unpublished. And it seems Bourbaki's shelves are piled high...

equal success, many of them were adopted by mathematicians in both France and other countries.

How did the international mathematicial community receive Bourbaki's treatise? Opinions vary from volume to volume. For example, among the most succesful are the books on general topology and on Lie groups and Lie algebras; in fact, Bourbaki's book on Lie Groups and Lie Algebras is still considered to be one of the best texts on this subject. The sales figures also show that the treatise as a whole was truly successful, and any self-respecting mathematics library in a university must own a full set of Bourbaki's treatise. In addition to English, *Éléments de mathématique* was translated into Russian and Japanese. Moreover, between 1950 and 1975, several volumes benefited from new editions with extensive revisions. Hermann certainly wouldn't have put so much effort into new editions if the first editions achieved only mixed success! According to Pierre Cartier, royalties from the book were equivalent to between fifty and seventy thousand current dollars per year, even during the dry years of the 1960s.

Some exercises from the book on topological vector spaces.
(For English translation, see appendix.)

Exercices. — 1) Donner un exemple d'une partie fermée et non compacte dans $\mathbf{R}^2$, dont l'enveloppe convexe ne soit pas fermée.

2) Dans l'espace $\mathbf{R}^n$, montrer que l'enveloppe convexe d'un ensemble compact est un ensemble compact (cf. § 1, exerc. 8).

¶ 3) Soit I l'intervalle $[0,1]$ de $\mathbf{R}$, et F l'espace vectoriel $\mathcal{C}(\mathrm{I})$ des fonctions numériques continues dans I. Soit E l'espace produit $\mathbf{R}^{\mathrm{F}}$ (ensemble des fonctions numériques finies définies dans F). Pour tout $a \in \mathrm{I}$, soit ε_a l'élément de E tel que $\varepsilon_a(f) = f(a)$ pour tout $f \in \mathrm{F}$.

a) Montrer que, lorsque x parcourt I, l'ensemble K formé des ε_x est compact dans E.

b) Soit μ l'élément de E tel que $\mu(f) = \int_0^1 f(t)dt$ pour tout $f \in \mathrm{F}$ (« mesure de Lebesgue »). Montrer que, dans E, μ est adhérent à l'enveloppe convexe de K, mais n'appartient pas à cette enveloppe (cf. *Fonct. var. réelle*, chap. II, § 1, prop. 5).

Who is Bourbaki For?

Paul Halmos (b. 1916), an American mathematician originally from Hungary, criticized the approach Bourbaki chose for its books.

Another indicator of Bourbaki's success is the opinions revealed in the numerous reviews appearing over the course of the years and of the new volumes of *Éléments de mathématique* in professional mathematical journals such as the famous *Mathematical Reviews*. Despite a few criticisms over technical points, most reviews were favorable. For example, the German algebraist Emil Artin writes in his 1953 review of *Algèbre* that "the connection between the different branches of mathematics becomes clearly visible," that Bourbaki tries "to present every concept with as much generality and abstraction as possible," and that the "terminology and notation were carefully developed and are being adopted by an increasing number of mathematicians." Artin also notes that the volumes on general topology were "already being used enthusiastically, especially by the youngest generation" and that while the presentation is "abstract, mercilessly abstract," anyone capable of overcoming the initial difficulties would see his efforts well rewarded. He concludes by emphasizing "the absolute success of the work." Alex Rosenberg echoes Artin's comments by writing in his 1960 report on a new chapter of the algebra book that the volume in question was satisfying and pleasant to read, and that he didn't even find the presentation to be "mercilessly abstract."

Yet not all reviews were flattering. For example, Edwin Hewitt wrote in 1956 that "the presentation is austere and monolithic. A huge number of definitions are strewn throughout the book, and many are left unmotivated. There is a continuous stream of exercises that are tiresome to complete. The reader must be prepared to refer constantly to the author's other works." Even those who praised the work's clear style and good exercises often doubted how useful the books would be for students. For example, E. Michael wrote in his 1963 review of a volume on general topology that "in fact, one wonders how many students really assimilate these chapters without having seen metric spaces elsewhere." There were more profound criticisms as well. The prominent mathematician Paul Halmos asserted in his 1953 report on the first four chapters of the book on integration that the answer to a student who asked whether the subject was important, the writing was clear, and the content was well-organized would be yes, but the answer to a professor who asked whether the book's perspective was the one that would best help a student understand the material and broaden his interests would be no. The books on set theory and on integration were the mostly heavily criticized, the books on set theory for their cavalier treatment of mathematical logic and the foundations of mathematics, and the books on integration for their peculiar approach to the subject.

What will become of Bourbaki's treatise in the twenty-first century? In principal, the publication of these books has not ceased, but it has significantly slowed down. Between 1980 and 1983, only five new or revised volumes were published. After 1983, the next installment of Bourbaki's treatise did not appear until 1998, when the most recent volume (the tenth chapter of the book on algebra) was published. Moreover, according to Pierre Cartier, only two or three hundred copies of this volume were printed, whereas the print runs of the 1960s were several thousand copies. Will Bourbaki publish again in the near future? Many doubt it (see Chapter 10), and some believe that any new books would be less significant than the earlier volumes. Indeed, Bourbaki's treatise has fallen victim to its own success. Many authors imitated Bourbaki's style, resulting

6. Propriétés des idéaux.

Dans ce n° et le suivant, nous ne parlerons que des idéaux à gauche ; nous laissons au lecteur le soin d'énoncer les propositions correspondantes pour les idéaux à droite et les idéaux bilatères.

Soit A un anneau, $\mathfrak{a}$ un idéal à gauche de A, B un sous-anneau de A ; l'intersection $B \cap \mathfrak{a}$ est un idéal à gauche dans l'anneau B. En particulier, si $\mathfrak{a} \subset B$, $\mathfrak{a}$ est un idéal à gauche dans B ; mais inversement, un idéal à gauche *dans* B n'est pas nécessairement idéal à gauche *dans* A.

Les idéaux à gauche de A étant identiques aux sous-groupes stables relatifs à une structure de groupe à opérateurs sur A, toutes les propriétés des sous-groupes stables d'un groupe à opérateurs (§ 6, n° 10) leur sont applicables. C'est ainsi que l'intersection d'une famille $(\mathfrak{a}_\iota)$ d'idéaux à gauche est un idéal à gauche ; parmi les idéaux à gauche qui contiennent une partie donnée M de A, il en existe un plus petit, qu'on appelle l'idéal à gauche *engendré* par M ; on dit aussi que M est un *système de générateurs* de cet idéal.

An extract from the book on algebra.
(For English translation, see appendix.)

in excellent books in almost every area of mathematics. It would be hard for any new books to be as original as the first. Yet Arnaud Beauville, a member of Bourbaki until 1997, said that "I can assure you that Bourbaki is working on some truly interesting books that would make real contributions to mathematics." Yet Bourbaki has not managed to complete these works, apparently due to lack of time.

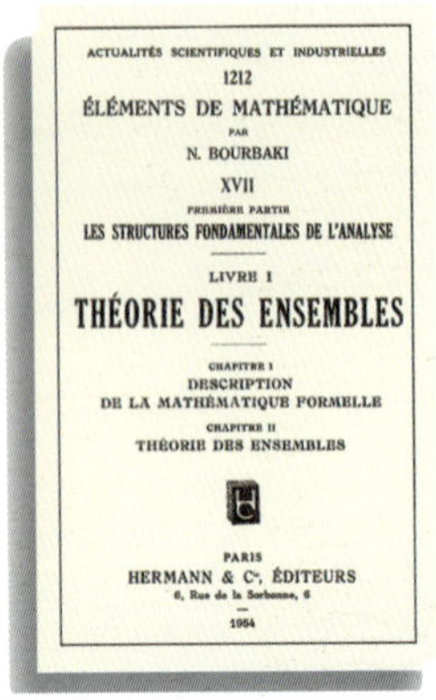
ACTUALITÉS SCIENTIFIQUES ET INDUSTRIELLES
1212
ÉLÉMENTS DE MATHÉMATIQUE
PAR
N. BOURBAKI
XVII
PREMIÈRE PARTIE
LES STRUCTURES FONDAMENTALES DE L'ANALYSE
LIVRE I
THÉORIE DES ENSEMBLES
CHAPITRE I
DESCRIPTION
DE LA MATHÉMATIQUE FORMELLE
CHAPITRE II
THÉORIE DES ENSEMBLES
PARIS
HERMANN & Cie, ÉDITEURS
6, Rue de la Sorbonne, 6
1954

2. Set Theory

Sets of points, sets of lines, sets of numbers, sets of functions,...the concept of a set is intuitively clear and is used in every branch of mathematics. However, certain difficulties arise when one considers sets with an infinite number of elements or when one tries to build set theory (which was created by Cantor at the end of the nineteenth century) on a completely rigorous and consistent foundation.

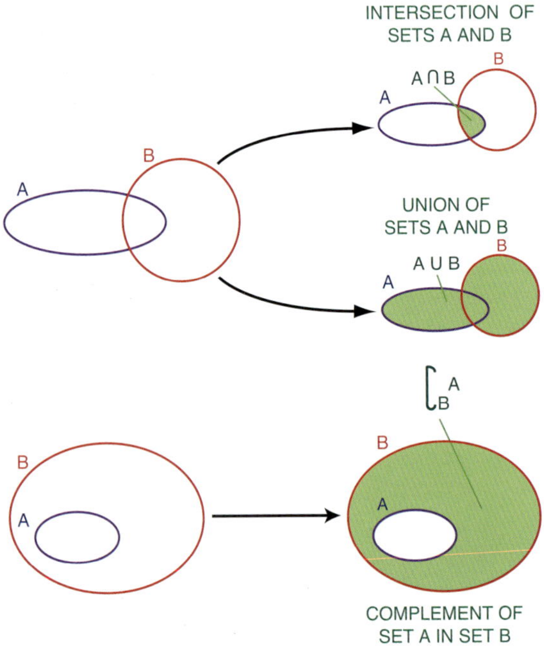

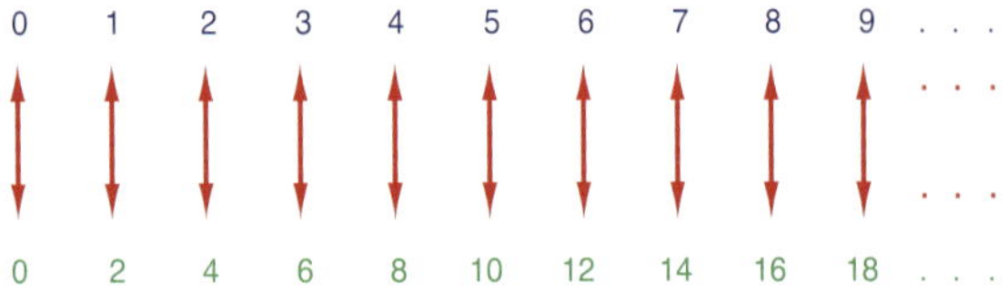

Elementary concepts of set theory include the union and intersection of two given sets and the complement of one set contained in another (see the figure opposite). To compare the size of two sets containing infinitely many elements, Cantor introduced the fundamental concepts of equipotence and cardinality. Two sets A and B are called equipotent if there is a bijection between the two sets (that is, a function such that each element of A corresponds to exactly one element of B and that each element of B corresponds to exactly one element of A). If A and B are equipotent, they are said to have the same cardinality; intuitively, the cardinality of a set represents the number of elements in that set. Infinite sets have infinite cardinalities, and this leads to counterintuitive results. For example, the set $\mathbf{N}$ of positive integers and the set of positive even integers are equipotent (see the diagram above), which roughly means that there are as many positive even integers as positive integers. While Arab mathematicians and Galileo already knew that fact, it was only Cantor who discovered and proved that the set $\mathbf{N}$ has the same cardinality as the set $\mathbf{Q}$ of rational numbers (numbers of the form p/q, where p and q are integers). Denoted by $\aleph_0$ ($\aleph$, aleph, is the first letter of the Hebrew alphabet), this is the cardinality of countably infinite sets. On the other hand, as Cantor proved, the set **R** of real numbers is not countable. The cardinality of **R** is greater than $\aleph_0$, which means that there are infinitely more real numbers than integers or rational numbers (this is even true of just the set of real numbers lying in the interval $[0, 1]$). But it can be shown that there are "as many" points on one side of a square as on the entire surface of the square!

While such properties are surprising, they are not paradoxical. Around 1900, however, mathematicians discovered true paradoxes in Cantor's naive set theory. One famous example consists of considering the set E of all sets that do not contain themselves; that is, E is the set of all sets X such that $X \notin E$, where $\in$ means "is an element of" and $\notin$ means "is not an element of." Is E an element of itself? If $E \in E$, then the definition of E implies that $E \notin E$. On the other hand, if $E \notin E$, then the definition of E implies that $E \in E$. Either case leads to a contradiction!

Logicians and mathematicians in the early twentieth century worked hard to build a more rigorous set theory founded on formal logic and devoid of contradictions. One result of this work is axiomatic set theory, which is based on a set of axioms known as the Zermelo-Fraenkel axioms and which is accepted by nearly all mathematicians. Bourbaki's book discusses a similar axiomatic set theory based on a slightly different set of axioms. Set theory is closely tied to the logical foundations of mathematics, an area of mathematics that is still active today.

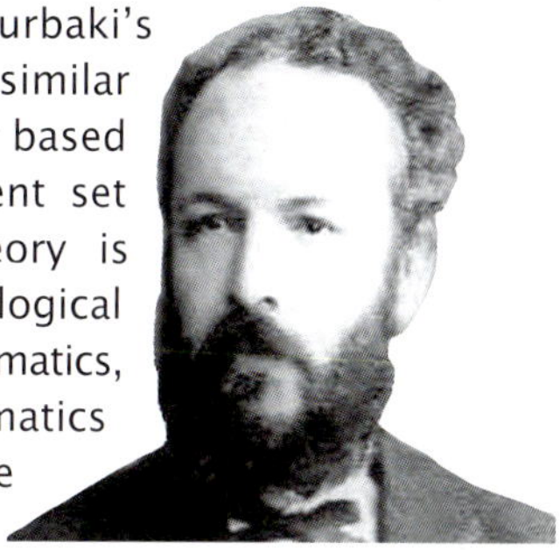

Georg Cantor, the father of set theory.

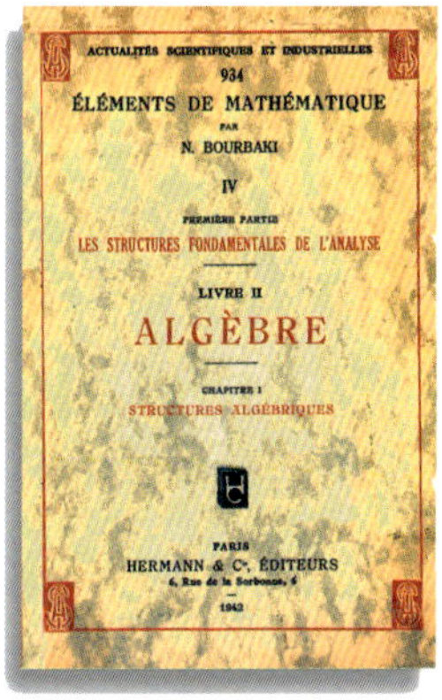

3. Algebra

As Bourbaki writes in the introduction to the group's book on Algebra, "Doing algebra is essentially performing calculations, that is, carrying out on the elements of a set the 'algebraic operations,' of which the most well-known are the 'four rules' of elementary arithmetic."

Algebra is an immense branch of mathematics that in some sense generalizes arithmetic to objects more abstract than common numbers. A well-known generalization of this type is the transition from concrete numbers to symbols representing an arbitrary number. For example, the identity $(a+b)^2 = a^2 + 2ab + b^2$, where a and b represent any two real numbers, is an algebraic identity.

Until the nineteenth century, algebra consisted mainly of performing such manipulations to solve algebraic equations, such as the general third-degree polynomial equation ($ax^3 + bx^2 + cx + d = 0$, where a, b, c, and d are given coefficients and x is unknown). Questions of this type led mathematicians to study and manipulate more complicated objects. For example, Évariste Galois (1811–1832) studied transformations consisting of permuting the roots of an equation. Since applying one permutation p_1 followed by a second p_2 is the same as applying a certain permutation $p_3 = p_2 \circ p_1$, called the composition of the permutations p_1 and p_2, the study of permutations led to the study of the general properties of the operation $\circ$. This, in turn, led to the definition of a general

Évariste Galois

structure called a group (see page 76), and the permutations form a group of transformations. In this way, algebra gained abstract structures like groups, rings, fields, and vector spaces, each defined by certain properties and computational rules. In addition, each abstract structure can materialize in many different ways. For example, it is not only the permutations that form a group; the integers equipped with the usual law of addition, for example, also satisfy the group properties. These algebraic structures are so general that they are found in almost every branch of mathematics and even in some other sciences. This is why many mathematicians, including Bourbaki, are so interested in thoroughly analyzing such structures.

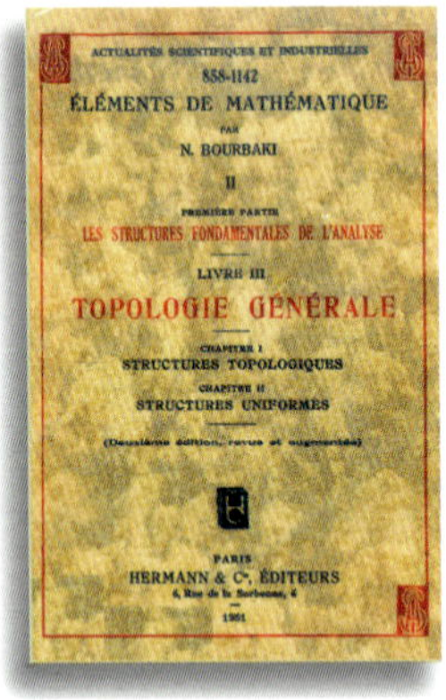

4. General Topology

Topology, known by the Latin *Analysis situs* during its prehistory, was not really born until the nineteenth century. The principal founder of this field was Felix Hausdorff (1868–1942), a mathematician as well as a poet and essayist, who committed suicide with his wife and sister-in-law when faced with deportation by the Nazis. Topology is fundamental to all mathematics and studies concepts of analysis in a general and abstract way. Topology deals with everything based on the concepts of limits, continuity, and neighborhoods. These concepts, initially used intuitively, were first defined rigorously as concepts of analysis (for example, the limit of an infinite sequence of numbers, the continuity of a function at a point or on an interval, and so forth). These definitions rely on the concept of a distance between two numbers x and y, measured, for example, by the absolute value $|x-y|$ of their difference. However, the concepts of limits, continuity, neighborhoods, and so forth can be defined without relying on the existence of a distance function. It is possible to define these concepts abstractly for sets satisfying certain properties; such spaces are called topological spaces, and topological spaces equipped with a distance function, which are particularly important objects, are called metric spaces (see pages 89 and 90).

When it deals with geometric objects, topology is the study of the properties of a surface that remain unchanged when the surface is continuously deformed, without creasing or tearing it. From a topological perspective, the surface of a sphere is equivalent to the surface of an egg or a sausage. On the other hand, the surface of a tire is not topologically equivalent to a sphere, but it is topologically equivalent to the surface of a coffee cup. Another example of a geometric problem in topology (not yet solved completely) is how to determine whether two knots are equivalent to one another. Some theorems in topology are surprising. One example is the important theorem known as the Brouwer fixed point theorem, which implies that it is impossible to comb a hairy ball so that none of the hairs stick out.

Another important branch of topology is called algebraic topology (which is not discussed in Bourbaki's book on general topology). Roughly, the goal of algebraic topology is to classify different types of topological spaces by characterizing each class with one or

Felix Hausdorff

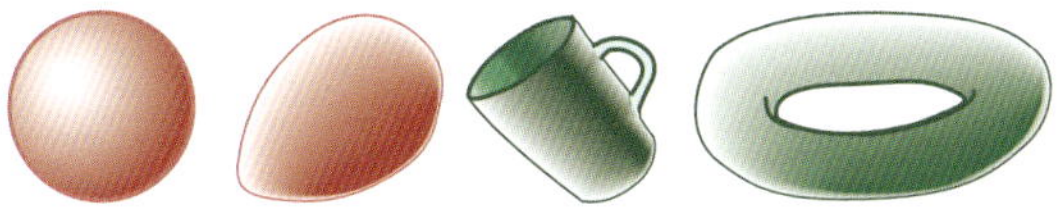

more algebraic objects (such as a number or a group), called invariants, which are the same for every topological space belonging to the same class.

5. Functions of One Real Variable

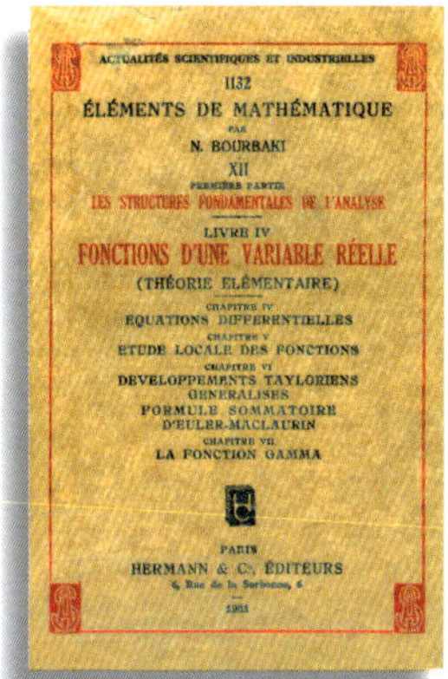

ACTUALITÉS SCIENTIFIQUES ET INDUSTRIELLES
1132
ÉLÉMENTS DE MATHÉMATIQUE
PAR
N. BOURBAKI
XII
PREMIÈRE PARTIE
LES STRUCTURES FONDAMENTALES DE L'ANALYSE
LIVRE IV
FONCTIONS D'UNE VARIABLE RÉELLE
(THÉORIE ÉLÉMENTAIRE)
CHAPITRE IV
ÉQUATIONS DIFFÉRENTIELLES
CHAPITRE V
ÉTUDE LOCALE DES FONCTIONS
CHAPITRE VI
DÉVELOPPEMENTS TAYLORIENS GÉNÉRALISÉS
FORMULE SOMMATOIRE D'EULER-MACLAURIN
CHAPITRE VII
LA FONCTION GAMMA
PARIS
HERMANN & Cie, ÉDITEURS
6, Rue de la Sorbonne, 6

Of all of Bourbaki's books, this one contains the most classical content. The book provides a detailed presentation of the subjects of classical analysis that deal with functions of one real variable. Functions of this type are numerous and even high school students study them; one of infinitely many examples of such functions is the function f defined by $f(x) = 3x + \sin x$, where x is any real number. But whereas all such functions studied in high school take real values, Bourbaki's book discusses such functions taking values in more general sets. More precisely, Bourbaki discusses functions $f : \mathbf{R} \to \mathbf{E}$, where E is a topological vector space over $\mathbf{R}$ (see box 6). A simple example of this sort of function is the function f that associates to each real number t the point (x, y) of the plane the coordinates $x = \cos t$ and $y = \sin t$. This function represents the trajectory of a point that moves with a constant speed of 1 along the circle with radius 1 and center at the origin.

The theory of functions of one real variable consists primarily of defining and analyzing the concepts of differentiation, anti-differentiation and integration, elementary functions (such as exponentiation, logarithm, and sine), differential equations (where the question is to determine a function based on the relation between the function and its derivatives), and the expansions of functions into infinite series, as well as of developing tools for comparing the behavior of two functions. Mathematicians have thoroughly understood the various aspects of this theory for over a century.

6. Topological Vector Spaces

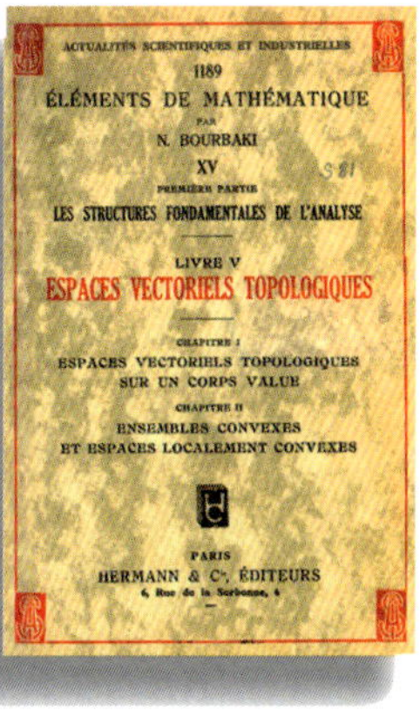

ACTUALITÉS SCIENTIFIQUES ET INDUSTRIELLES
1189
ÉLÉMENTS DE MATHÉMATIQUE
PAR
N. BOURBAKI
XV
PREMIÈRE PARTIE
LES STRUCTURES FONDAMENTALES DE L'ANALYSE
LIVRE V
ESPACES VECTORIELS TOPOLOGIQUES
CHAPITRE I
ESPACES VECTORIELS TOPOLOGIQUES SUR UN CORPS VALUÉ
CHAPITRE II
ENSEMBLES CONVEXES ET ESPACES LOCALEMENT CONVEXES
PARIS
HERMANN & Cie, ÉDITEURS
6, Rue de la Sorbonne, 6

Vector spaces are fundamental structures of linear algebra, a vast subject originating in the study of systems of linear equations and having a wide range of applications. A vector space is a set whose elements, called vectors, can be added to each other and multiplied by "numbers" called scalars. More precisely, if $\boldsymbol{v}$ and $\boldsymbol{w}$ are vectors, then $\boldsymbol{v}+\boldsymbol{w} = \boldsymbol{w} + \boldsymbol{v}$ is also a vector; and if α is a scalar and $\boldsymbol{v}$ is a vector, then $\alpha\boldsymbol{v}$ is also a vector. One of the most simple examples of a vector space is the set of vectors in the plane (each vector is represented by an arrow with a given length and direction), which are added by the parallelogram rule (see the diagram at the top of page 64) and can be multiplied by real numbers. This is called a vector space over $\mathbf{R}$, since in this case the scalars are real numbers. One can also construct vector spaces over $\mathbf{C}$ (where the scalars are complex numbers), or, more generally, over an arbitrary field K (see page 81 for the definition of a field).

A topological vector space is a vector space equipped with a topological structure in which any vector can be continuously deformed into any other vector. In more precise language, a topological vector space is a vector space where the function that associates to each pair ($\boldsymbol{v}$, $\boldsymbol{w}$) of vectors to the vector $\boldsymbol{v}+\boldsymbol{w}$ and the function that associates to each scalar α and vector $\boldsymbol{v}$ the vector $\alpha\boldsymbol{v}$ are both continuous. The abstract and elaborate theory of topological vector spaces generalizes the theory of normed vector spaces, where each vector has a norm (which is a generalization of the concept of length). It is used in functional analysis, which is the study of infinite-dimensional spaces whose elements are functions.

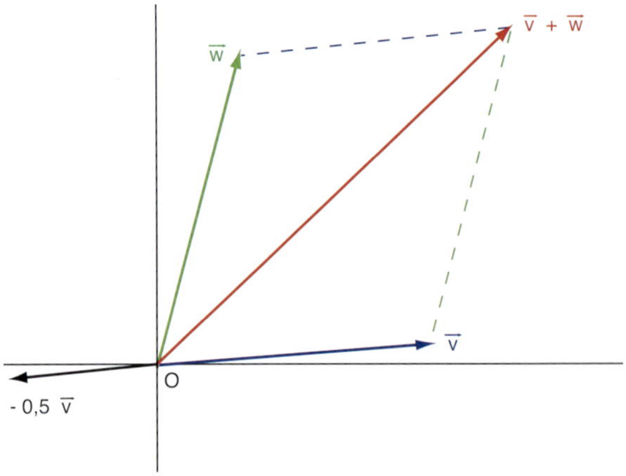

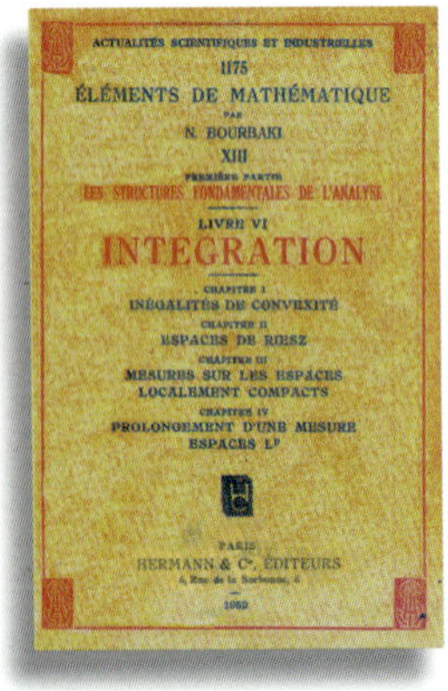

7. Integration

The concept of integration has numerous applications, both in mathematics and in other sciences. In elementary texts, integration is presented in terms of the antiderivative. Given a function f, one says that F is the antiderivative of f if for every x the derivative $F'(x)$ equals $f(x)$. Then the integral of f over an interval $[a,b]$ is defined by the formula $\int_a^b f(x)dx = F(b) - F(a)$.

Historically, however, the origins of integration lie in the calculation of areas and volume. By definition, the integral $\int_a^b f(x)dx$ is the area of the region in the x,y-plane bounded by the curve representing f, the x-axis, and the vertical lines $x=a$ and $x=b$. A somewhat more precise but still simplified definition is that the integral $\int_a^b f(x)dx$ is the limit as N approaches infinity of the sum $f(x_1)\Delta x + f(x_2)\Delta x + \cdots + f(x_N)\Delta x$, which corresponds to subdividing the interval $[a,b]$ into N little intervals of length $\Delta x = (b-a)/N$, and where x_i is a point in the ith interval. This is equivalent to calculating the area in question by approximating the region by narrower and

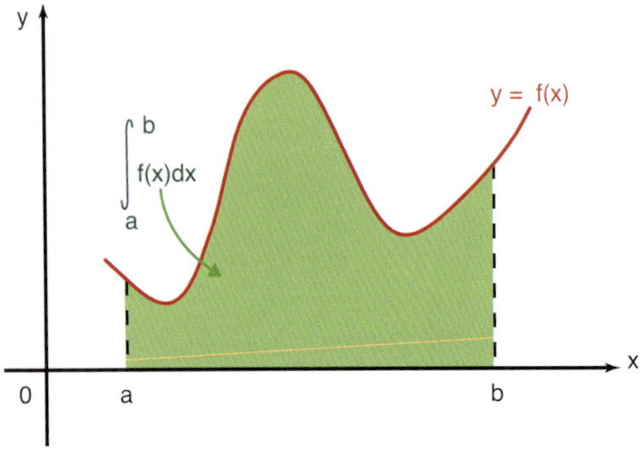

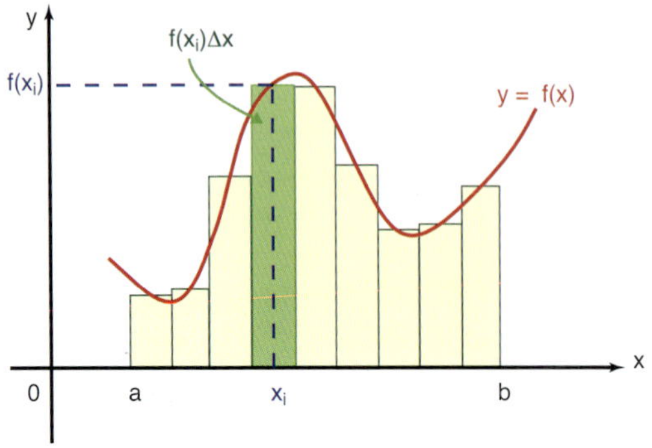

narrower rectangles (see the diagram). With this approach, the formula $\int_a^b f(x)dx = F(b) - F(a)$ becomes a theorem—although only valid under certain hypotheses—rather than a definition.

The concept of integration extends to more general functions (such as ones depending on more than one variable), and instead of the interval $[a, b]$, the domain of integration can be a subset of a plane, of three-dimensional space, of n-dimensional space, or even of a more abstract set.

Integrating a function over a given domain is in some sense adding the values $f(P)$ of the function over all the points P in the domain. But to obtain a finite result, it is necessary to multiply each $f(P)$ by something infinitely small, since there are infinitely many points P in a domain like an interval or a region of a plane. (This is the purpose of the Δx used above.) The goal of integration theory is to do this in the most rigorous and general way, so that the properties and theorems would hold for the most different types of functions as possible. In this respect, the theory that Henri Lebesgue built on Émile Borel's measure theory in the early 1900s marked a major step in the development of integration (see the box on page 127).

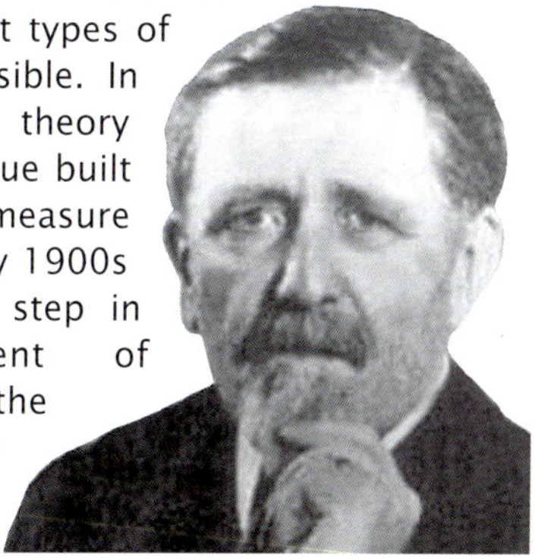
Émile Borel

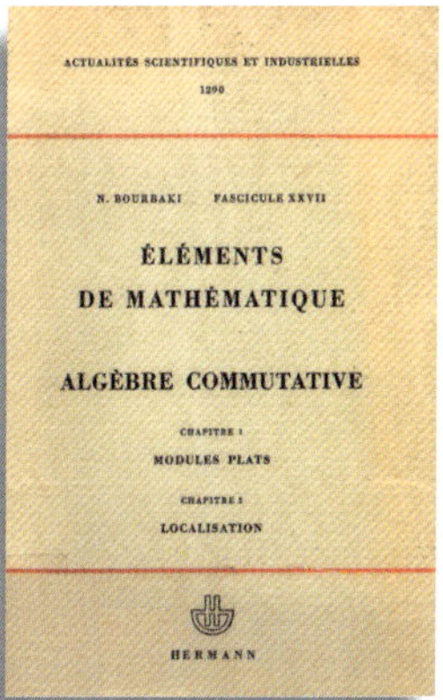

8. Commutative Algebra

A commutative, or abelian, operation on two elements of a set is simply an operation for which the result doesn't depend on the order of the two elements. For example, the usual addition and multiplication of real numbers are commutative since $a + b = b + a$ and $ab = ba$ for all real a and b. On the other hand, the composition of functions is not commutative, since in general $f \circ g$ does not equal $g \circ f$; in other words, for two functions f and g it is not generally the case that $f(g(x))$ equals $g(f(x))$ for all x.

Although the concept of commutativity is simple, Bourbaki's book on commutative algebra is not. The work discusses very technical questions "that appeared over the course of the development of algebraic number theory and, later, of algebraic geometry." Recall that algebraic geometry is based on the study of solutions to algebraic equations, which can be interpreted as geometric objects (as, for example, the equation $x^2 + y^2 = 1$ can be interpreted as a circle with radius 1). To study the properties of these geometric objects, one approach is to directly study the solutions of the equations. This is the approach of traditional geometry. Another approach is to work with just the equations. For example, one fundamental problem is to determine whether two given systems of equations represent the same geometric object without actually solving the equations. Commutative algebra uses techniques of transforming equations to answer such questions.

More specifically, Bourbaki's book presents "concepts that can, in principal, be applied to any commutative ring and to modules over these rings." It would be impossible to explain what the book covers in more detail, but we can describe what rings and modules are. A ring is a set whose elements can be added and multiplied (see page 78 for a precise definition); for example, the set **Z** of integers is a ring. The addition law in a ring is, by definition, always commutative. When the multiplication law is also commutative (as is the case for **Z**), the ring is called a commutative ring.

The concept of modules is a generalization of the concept of a vector space (see box 6). In a vector space, the vectors can be multiplied by scalars belonging to a field K (often the field **R** of real numbers of the field **C** of complex numbers). When the scalars are taken from a ring R instead of a field K, then the structure is a module over R rather than a vector space over K.

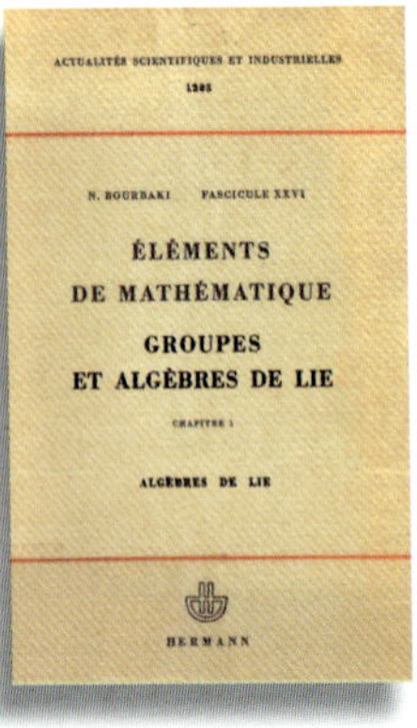

9. Lie Groups and Lie Algebras

A set G is called a group if it is equipped with an operation, denoted here by $*$, that satisfies the following three properties:

1. Associativity: for any x, y, z in G, $x * (y * z) = (x * y) * z$.
2. Existence of an identity element: there exists an element e of G such that $e * x = x * e = x$ for all x in G.
3. Existence of an inverse for any element: for each x in G, there exists an element x^{-1} such that $x * x^{-1} = x^{-1} * x = e$.

Such structures are found everywhere. For example, it is easy to check that the set of numbers $+1$ and -1 form a two-element group when $*$ is taken to be the usual law of multiplication. Similarly, the set of translations of the plane, equipped with the operation of composition, forms a group with infinitely many elements.

A Lie group, named after the Norwegian mathematician Sophus Lie (1842–1899), is a particular type of group whose definition is much more complex. An elementary example of a Lie group is the set of complex numbers with modulus 1, which can be written as $e^{i\theta}$ (or, equivalently, as $\cos\theta + i\sin\theta$), equipped with the usual operation of multiplication. The identity element is $1 = e^{i0}$ and the inverse of $e^{i\theta}$ is $e^{-i\theta}$. This group can be interpreted geometrically as the group of rotations of the plane about the origin, since if z is a complex number, then $z' = e^{i\theta}z$ has the same modulus as z and has argument $\arg(z') = \arg(z) + \theta$ (in other words, the point representing z' is obtained by rotating the point that represents z about the origin by an angle of θ).

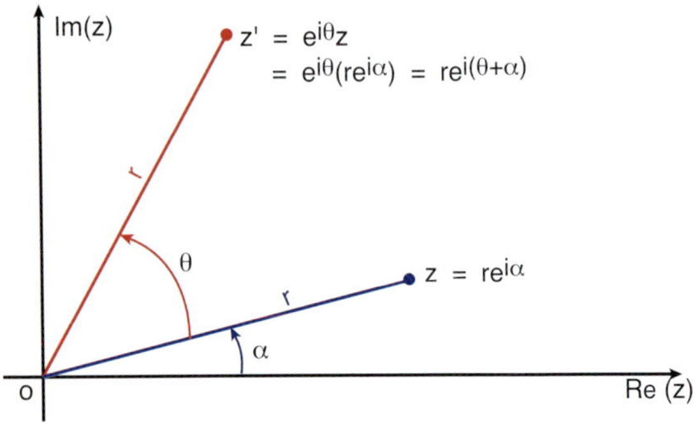

Very roughly, a Lie group is a group containing infinitely many elements that can be labelled by one or more parameters. In the case of a single parameter θ, each value of θ defines an element $g(\theta)$ of the group G, and one element can be transformed into another by continuously varying θ. Other conditions must also be satisfied for a group to be a Lie group (in particular, the set the parameter belongs to must by an analytic variety), but we won't discuss them here.

What you should remember about Lie groups is that they have very rich properties and play an important role in mathematics (particularly in the study of differential equations and differential geometry), as well as in physics and chemistry (for example, in the theory of elementary

Sophus Lie

particles and the classification of energy states of atoms and molecules). The analysis of Lie groups gives rise to underlying structures called Lie algebras, which are also studied in their own right (an algebra is a vector space whose vectors can be multiplied, and a Lie algebra is an algebra that satisfies certain properties).

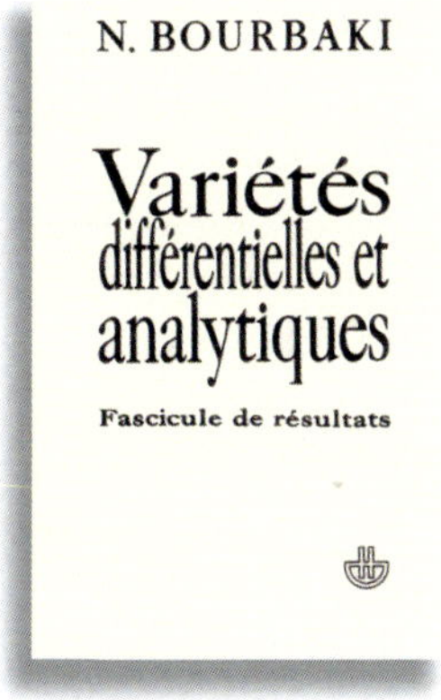

10. Differential and Analytic Varieties

A differential or analytic variety is a mathematical object that generalizes the concepts of smooth curves and surfaces.

One of the main properties of a differential or analytic variety is that a local system of cartesian coordinates can be attached to any point of the variety. For example, consider a curve in the plane and a point P on the curve. There is a line tangent to the curve at P, and this tangent almost coincides with the curve at the points very close to P. Thus one can identify points on the curve close to P with x-coordinates taken along the tangent at P. Since in this case one local coordinate is needed to identify each point, the curve is called a one-dimensional differential variety. It is embedded in two-dimensional Euclidean space (the Euclidean plane). Similarly, a surface, such as the usual sphere, is a two-dimensional variety, since in this case the tangent line is replaced by a tangent plane and so two coordinates are needed to identify a point near P. It is embedded in three-dimensional Euclidean space.

More generally, it is possible to imagine (but not draw) varieties of dimension n, where n is any integer. Moreover, varieties do not have to be embedded in a space of higher dimension; instead, they can be defined intrinsically without referring to any space around it. To visualize this, imagine a blind but intelligent ant that lives on some surface. Although it is not aware that the surface is actually in three-dimensional space, the ant can determine the surface's geometric properties—for example, the curvature at each point—simply by walking along the surface. In fact, humans are in this very situation. Einstein's theory of general relativity views our universe as an intrinsic four-dimensional variety, with three dimensions for space and one for time. While it appears flat on a very small scale, space-time is actually a curved variety.

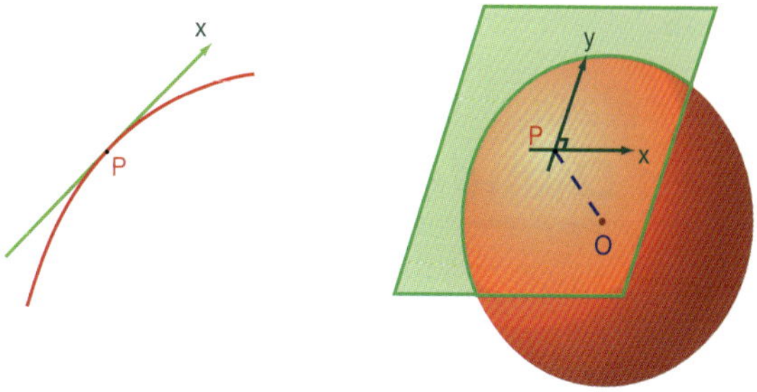

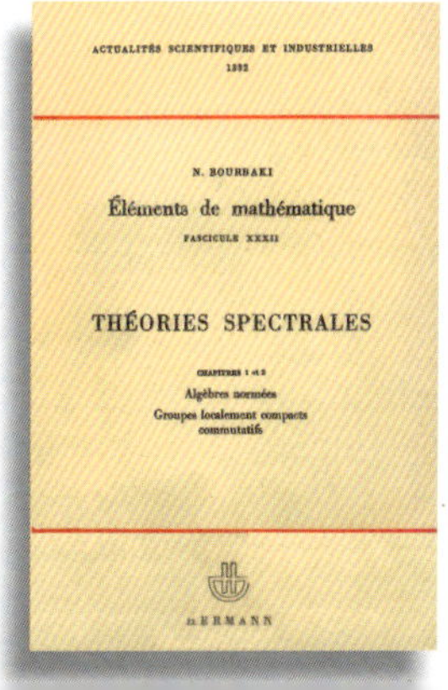

11. Spectral Theory

Spectral theory is an extension of the study of concepts called eigenvalues and eigenvectors. These are thoroughly understood in the case of finite dimensional vector spaces, which are vector spaces in which the vectors are determined by a finite number of coordinates. In infinite dimensional vector spaces (for example, the space of complex-valued functions $f : \mathbf{R} \to \mathbf{C}$ such that the integral of $|f|^2$ is a finite number, which has important applications in quantum physics and quantum chemistry), the behavior of eigenvalues and eigenvectors is much more complex and rich.

Questions about eigenvalues and eigenvectors rely on the concept of linear transformations acting on a vector space. So first, what is a linear transformation? A linear transformation is a map, let's call it M, that associates every vector $\boldsymbol{v}$ in the vector space V to another vector $\boldsymbol{w}= A(\boldsymbol{v})$ and that satisfies the property of linearity: $A(\alpha_1 \boldsymbol{v}_1+\alpha_2\boldsymbol{v}_2) = \alpha_1 A(\boldsymbol{v}_1) + \alpha_2 A(\boldsymbol{v}_2)$ for any vectors $\boldsymbol{v}_1$ and $\boldsymbol{v}_2$ and any scalars α_1 and α_2. Letting V be the set of vectors in the plane, it is easy to check that the rotation R by a given angle about the origin and the projection P onto the x-axis are linear transformations on V.

If there is a nonzero vector $\boldsymbol{v}$ and a scalar α such that $A(\boldsymbol{v}) = \alpha\boldsymbol{v}$, then $\boldsymbol{v}$ is called an eigenvector and α is called an eigenvalue. Here is a simple example (see the diagram opposite). Let P be projection onto the x-axis. Any horizontal vector is invariant under the action of P, that is $P(\boldsymbol{v}) = \boldsymbol{v}$ for any $\boldsymbol{v}$ in V. Therefore, every horizontal vector is an eigenvector with eigenvalue 1. Similarly, any vertical vector $\boldsymbol{v}$ satisfies $P(\boldsymbol{v}) = \boldsymbol{0} = 0\cdot\boldsymbol{v}$, so any vertical vector is an eigenvector with eigenvalue 0. These are the only two eigenvalues of the projection P.

The name "spectral theory" comes from an object called a spectrum. In many cases (including the case of finite dimensional vector spaces), the spectrum of a linear transformation is the same as the set of the transformation's eigenvalues. Studying the eigenvalues and eigenvectors of a linear transformation is important because these help characterize the properties of the transformation and considerably simplify certain calculations.

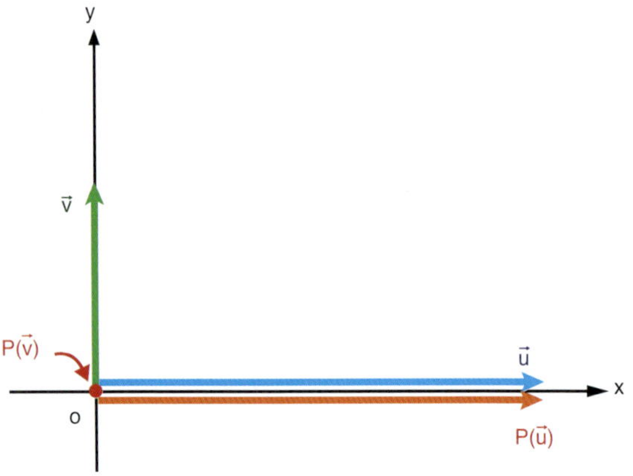

Jean Dieudonné (1906–1992)

Jean Dieudonné was born in Lille on July 1, 1906. His father worked in the textile industry and his mother was a teacher. In 1915, the German occupation of Lille forced him to be evacuated to Switzerland with his mother and sister. From Switzerland they reached Paris to join his father, who had been mobilized the year before. There Jean entered the Lycée Condorcet. His father sent him to the Isle of Wight in England for the school year 1919–1920 to study English; it was there that he started to like mathematics. Upon returning to France, he entered the Lycée Faidherbe in Lille. He graduated in 1923, obtaining his baccalaureat as well as winning first place in mathematics in an annual mathematics competition. He entered l'École Normale Supérieure in 1924. He passed the teacher qualification examinations in 1927 with flying colors, placing first and impressing the panel of examiners.

Upon completing his military service, Jean Dieudonné obtained a fellowship to spend one year at Princeton University. He then returned to l'École Normale Supérieure, where he spent the academic year 1929–1930 before obtaining a fellowship from the Rockefeller Foundation to spend a few months at the University of Berlin (with Ludwig Bieberbach) and the University of

Zurich (with George Pólya). In 1931 he completed his thesis, *Recherches sur quelques problèmes relatifs aux polynômes et aux fonctions bornées d'une variable complexe*, which dealt with functions of a complex variable. He became a lecturer in the Faculty of Sciences at Bourdeaux and then in the Faculty of Sciences at Rennes, where he stayed from 1933 until 1937. According to Dieudonné, the two most important events of his life occurred in August 1934. First, he met his future wife, Odette Clavel, in Pleyel Hall at the Concerts Lamoureux (Dieudonné loved music and played piano every day). Second, the Bourbaki group was formed.

In 1937, Jean Dieudonné was hired as an assistant professor at the Faculty of Sciences at Nancy and subsequently promoted to full professor. He spent two years (1946–1948) at the University of São Paulo in Brazil and worked in the United States from 1952 to 1959. He returned to France to work at the Institut des Hautes Études Scientifiques in Bures-sur-Yvette (near Paris). He stayed there until 1964, when he became the dean of the brand-new Faculty of Sciences at Nice. In 1970 he retired to Paris and devoted himself to the history of mathematics. He died on November 29, 1992.

Dieudonné's work dealt with algebra, topological vector spaces, topology, and Lie groups. One of his innovations was the concept of paracompact spaces, which is used in algebraic topology. Thanks to his vivid personality, vast knowledge, good memory, and immense ability to work and write, Dieudonné was a powerhouse in Bourbaki until his retirement from the group in 1956. He wrote many texts, including textbooks and books on the history of mathematics, in which he expressed his Bourbakian views. He helped coordinate the mathematical articles in the *Encyclopaedia Universalis*, which were well-written but relatively technical. While working at the IHES, Dieudonné helped Alexandre Grothendieck write his *Éléments de géométrie algébrique*. Another important example of his writing is *Pour l'honneur de l'esprit humain - les mathématiques aujourd'hui* (published in English as *Mathematics - the Music of Reason*), which was written for the general public and published in 1987.

Dieudonné at the blackboard during a Bourbaki conference in 1951.

5

Towards Axioms and Structures

An intellectual construction endowed with profound unity, a hierarchy of abstract structures built on a foundation of axioms: this is how Bourbaki viewed mathematics. The group's beliefs gained many followers.

Nicolas Bourbaki did much more than write a voluminous and ambitious mathematical treatise. During its golden years of the sixties and seventies, the group also propagated a certain vision of mathematics. This new ideology was adopted by many mathematicians around the world, and it even influenced some fields outside of mathematics. Remember that Bourbaki's global views of mathematics developed over the course of the group's first ten of fifteen

The ritual tea hour at the Institut des Hautes Études Scientifiques at four o'clock. In the foreground are Jean Dieudonné, Alexandre Grothendieck (from behind), Michel Demazure, and François Bruhat.

years, as its members completed the outline of their treatise and started to write the first volumes. Also remember that the views held by Bourbaki as a group are not necessarily the same as the those held by its individual members. While some of the Bourbaki members propagated the Bourbakian ideology quite faithfully, Bourbaki is heterogeneous like any group of humans. Opinions varied significantly from one member to another, and it is a job for historians to sort out these various opinions. Given the secrecy and collective nature of Bourbaki, this job remains a hard one to this day. Finally, it is important to remember that the times and the members have changed, and so the Bourbakian vision of today is probably not the same as that of the fifties. However, as the Bourbaki of today does not express its vision of mathematics (if it has one at all), we must settle for studying the vision held in the past.

Bourbaki spread its conception of mathematics through various texts and lectures. One of the most important of these is the article *L'architecture des mathématiques* ("The Structure of Mathematics") published in 1947 and signed by Nicolas Bourbaki himself. Although this article is a real manifesto of Bourbaki's ideology, it appears that the group didn't discuss the article much before its publication. *L'architecture des mathématiques* was probably written by Dieudonné, who usually didn't express himself subtly. Discussing this article in his book *Mathématique: (récit)*, the author-mathematician Jacques Roubaud writes that "Bourbaki calmy wields Neanderthalian clubs of philosophy, in contrast to his habitual rabbit-like caution."

So how did Bourbaki see mathematics? Its philosophy revolves around three key notions: the unity of mathematics, the axiomatic method, and the study of structures. Today the unity of mathematics has become a well-worn topic that comes up every time mathematicians take a slightly global look at mathematics. Mathematicians are always emphasizing that geometry, algebra, analysis, and number theory are no longer separate topics; that modern mathematical research applies to all domains; that, for example, the proof of a theorem in number theory uses a mixture of concepts and methods from analysis, geometry, and algebra. Yet this unity was not so blatantly obvious to mathematicians of the thirties and forties, the era when Bourbaki formed. In *L'architecture des mathématiques*, Bourbaki writes that "one can ask himself whether this extravagant proliferation [of mathematical results] is the product of a vigorous organism that acquires more cohesion and unity with every addition it receives, or whether, on the contrary, if it is the outwards manifestation of an increasing fragmentation due to the very nature of mathematics, and whether mathematics is becoming a Tower of Babel of autonomous disciplines,

Euclid developed a precursor to the axiomatic method.

isolated from one another in their goals, methods, and even in their language. In short, is modern mathematics a single *mathématique* or many *mathématiques*?"

Bourbaki's reply to this question is clearly in favor of the singular *mathématique*, which the group uses in the title of its treatise, *Éléments de mathématique*. In the eyes of Bourbaki, this unity is strong, in some respects even stronger than it is considered today. As Bourbaki writes, "We believe that the internal evolution of mathematical sciences, despite appearances, has strengthened the unity of its various subjects more than ever. It has created a sort of nucleus that is more cohesive than ever before. The most important aspect of this evolution is the systematization of the connections between the various areas of mathematics. This is known as the axiomatic method."

A stamp and portrait of Richard Dedekind (1831–1916). Dekedind was one of the fathers of modern algebra.

The Axiomatic Method According to Hilbert

What exactly is the axiomatic method? To begin with, an axiom is often a self-evident principle whose validity is accepted without proof. An axiom is a sort of fundamental truth, but an axiom can also be an invented property or rule that has no immediate connection to reality. An axiomatic theory starts with the definitions of the objects it will deal with. Then axioms (sometimes called postulates) that the objects in question must obey are added. Finally, logical chains of reasoning, whose validity can be verified without reference to intuition or experience, lead from the objects and axioms to less obvious properties, called theorems.

The most ancient example of the axiomatic method is found in Euclid's geometry. In his *Elements*, Euclid of Alexandria presents geometry by starting with the fundamental objects of a point ("that which has no part"), a line ("a length without thickness"), a curve, a plane, and so forth. He proceeds by stating five axioms: 1) Given any two points, a line segment exists that joins them; 2) A line segment can be extended indefinitely; 3) A circle can be constructed with any given center and radius; 4) All right angles are equal; and 5) If a line crosses any two lines such that the sum of the interior angles is less than two right angles, then the two lines cross when extended on the corresponding side. The fifth axiom is equivalent to the famous parallel postulate ("Given a line and point not on the line, there exists a unique line passing through the point and parallel to the first line"), which has made so many mathematical minds spin.

David Hilbert (1862–1943).

From Bourbaki to Oulipo, Lévi-Strauss, and Piaget

In the fifties and sixties, Bourbaki's vision of mathematics influenced areas outside of mathematics. In particular, it expressed itself in literature with Oulipo (short for *Ouvroir de littérature potentielle*, meaning "Workshop for Potential Literature"), a movement related to surrealism. Created in 1960 by Raymond Queneau and François le Lionnais, Oulipo aimed to explore new forms of literature obtained by imposing constraints ("structures") of a mathematical nature. For example, Queneau wrote a literary translation of Hilbert's *Foundations of Geometry*, which he titled *Les fondements de la littérature d'après David Hilbert* ("The Foundations of literature according to David Hilbert"). In this work, Queneau imitates the axioms of geometry developed by Hilbert by replacing the words "points," "lines," and "planes" by "words," "sentences," and "paragraphs," respectively. The results of such exercises are often hilarious. Oulipo, which included some mathematicians (notably Jacques Roubaud and Claude Berge), had direct contacts with Bourbaki. For example, Queneau attended a Bourbaki conference in 1962. Their humor, taste for secrecy, and use of structures helped bring the two groups together.

Raymond Queneau (1903-1976) in 1951.

Bourbaki's structuralism was also fashionable in French social sciences. Structuralism had been developing in the social sciences since the fifties, following the publication of *Structures élémentaires* by the anthropologist Claude Lévi-Strauss. Lévi-Strauss had met André Weil in 1943 in New York, which led to a small collaboration: by using group theory, André Weil solved a combinatorial problem about marriage rules in an Australian tribe. This contribution appeared in an appendix of Lévi-Strauss's book. Another cross-disciplinary encounter was that of Jean Dieudonné with Jean Piaget during a 1952 conference on mathematical and mental structures. This led Piaget to consider the existence of a direct connection between the structures in a child's mental processes and the mathematical mother-structures.

While Bourbaki's structures were often mentioned in social science conferences and publications of the era, it seems that they didn't play a real role in the development of these disciplines. David Aubin, a science historian who analyzed Bourbaki's role in the structuralist movement in France, believes Bourbaki's role was that of a "cultural connector." According to Aubin, while Bourbaki didn't have any mission outside of mathematics, the group represented a sort of link between the various cultural movements of the time. Bourbaki provided a simple and relatively precise definition of concepts and structures, which philosophers and social sciences believed was fundamental both within their disciplines and in bridges among different areas of knowledge. Despite the superficial nature of these links, the various schools of structuralist thinking, including Bourbaki, were able to support each other. So, it is not a coincidence that these schools suffered a simultaneous decline in the late 1960s.

An extract from *Les fondements de la littérature d'après Hilbert* by Raymond Queneau. Published in *La bibliothèque Oulipienne, présenté par Jacques Roubaud* (Slatkine).

Group I (Incidence Axioms)

I, 1 *There exists a sentence containing any two given words.*

Commentary: Obvious. Example: Consider the two words "A" and "A." There is a sentence containing these two words: "A violinist gave an A to the singer."

I, 2 *There exists at most one sentence containing any two given words.*

Commentary: This, on the other hand, is surprising.

But let's consider two words such as "long" and "bed." It is clear that once the sentence containing them has been written, namely, "For a long time I would go to bed early," any other expression like "for a long time I would retire to bed early" or "for a long time I never went to bed late" is a pseudo-sentence that must be rejected in accordance with this axiom.

Scholium: Naturally, if one wrote, "For a long time I would retire to bed early," it is the sentence, "For a long time I would go to bed early" that must be rejected in accordance with axiom I.2. In other words, *In Search of Lost Time* cannot be written twice[1].

I, 3 *Any sentence contains at least two words. There exist at least three words not all belonging to the same sentence.*

Commentary: Thus there are no sentences consisting of a single word. "Yes," "No," "Hey," and "Psst" are not sentences. As for the second part of the axiom: one thus supposes that the language in question contains at least three words (this is trivial for the case of French) and that no sentence can contain *all* the words of the language (or all but one, or all but two).

1, 4a *There exists a paragraph containing three given words not all belonging to the same sentence.*

Commentary: This immediately implies that a paragraph must contain at least two sentences.

One will note that the statements of axioms I, 1 through I, 4 do not comply to axiom I,2 because all four require the words "words" and "sentences" for their formulation, while this axiom states that there can only be a single sentence containing these two words.

This yields the following axiom of metaliterature: *Axioms do not obey the axioms.*

1, 4b *Every paragraph contains at least one word.*

Commentary: "Yes," "No," "Hey," and "Psst," which are not sentences according to I, 3, do not constitute paragraphs by themselves.

I, 5 *There exists at most one paragraph containing three given words not all belonging to the same sentence.*

Commentary: Like I, 2, this axiom is an axiom of uniqueness, the uniqueness of paragraphs in this case. In other words, if three words not all belonging to the same sentence are written in one paragraph, one cannot reuse these words in another paragraph. But, one might protest, what if they all belonged to the same sentence in the second paragraph? This axiom excludes this possibility.

I, 6 *If two words in a sentence belong to a paragraph, then all the words in that sentence belong to that paragraph.*

Commentary: No commentary necessary. [...]

[1] *Note that the opening line of Swann's Way, the first volume of Marcel Proust's In Search of Lost Time (Trans. C.K. Scott Moncrieff and Terence Kilmartin), is "For a long time I would go to bed early." Translator.*

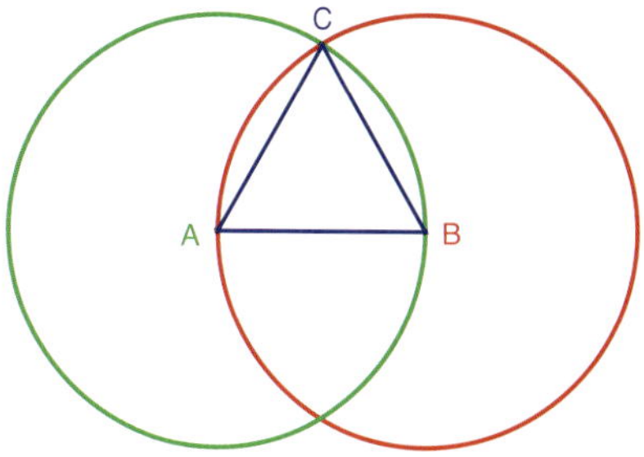

When A and B are the centers of the circles, the triangle ABC is equilateral.

The important thing here is not the content of Euclid's axioms but the fact that Euclid founded a theory on axioms that he could use (or more accurately, believed he could use) to derive geometric constructions and properties, such as the Pythagorean Theorem, rigorously. In reality, Euclid did not always use flawless rigor. He unwittingly used many properties that were suggested by visual intuition rather than included in the axioms or proved rigorously. A perfect example of this is his first proposition, which consists of constructing an equilateral triangle with a given side AB. To do this, Euclid draws two circles of radius AB, one centered at A and the other centered at B. He then proves that triangle ABC, where C is one of the two points where the two circle intersect, is equilateral. However, to prove this Euclid uses the property that the two circles must intersect in at least one point. While this is visually obvious, Euclid should have justified this statement rigorously.

Despite this, what Euclid had developed was an important precursor to the axiomatic method. Moreover, it is not at all easy to come up with a coherent system of axioms for Euclidean geometry. In fact, this was achieved only a century ago, primarily by the great German mathematician David Hilbert in his book *Grundlagen der Geometrie* ("Foundations of Geometry"), published in 1899. Hilbert's project did far more than repair the flaws in Euclid's axioms: it was a stamp of approval for the modern axiomatic method in general.

What distinguishes the modern axiomatic method advocated by Bourbaki from Euclid's method? One main difference lies in its formal nature: the modern axiomatic method does not try to define basic concepts (like points and lines) that the theory in question will discuss. These basic concepts are treated as abstract entities whose nature and concrete meaning are insignificant. To illustrate this, Hilbert jokes that the "points," "lines," and "planes" used in the axioms of geometry could just as well be called "chairs," "tables," and "beer bottles." Only the relations between the fundamental entities defined by the axioms are important. The properties deduced from such a formal theory are completely general, since they could apply to a very different set of objects, as long as the axioms for that set of objects are the same. Later we'll see the simple example of how axioms are used to define a structure called a group.

It's important to understand that mathematicians do not construct sets of axioms directly out of nothing. The mathematician starts by studying a certain set of objects, and then he develops a set of axioms based on these objects. As Henri Cartan (one of the founders of Bourbaki) explained in a 1958 lecture in Germany, "A mathematician who sets about proving a theorem has in mind certain well-defined

1. What is a Group?

A non-empty set G is called a group if it is equipped with a composition law, denoted for example by $*$, that associates to each pair (x, y) of elements in G an element denoted by $x*y$ (also in G) and satisfies the following properties.

1. Associativity: for any x, y, z in G, $x * (y * z) = (x * y) * z$.
2. Existence of an identity element: there exists an element e of G such that $e * x = x * e = x$ for any x in G.
3. Existence of an inverse for any element: for each x in G, there exists an element x^{-1} such that $x * x^{-1} = x^{-1} * x = e$.

If the law $*$ is commutative—that is, if $x * y = y * x$ for any x and y in G—then we say that G is an abelian group. It is easy to deduce numerous properties from the above axioms. As an example, let's prove that if $x * y = x * z$, then $y = z$. Multiplying the both sides of the equality $x * y = x * z$ by x^{-1} on the left yields $x^{-1} * (x * y) = x^{-1} * (x * z)$. Then, by using axiom 1 followed by axiom 3, we obtain $e * y = e * z$. Finally, axiom 2 implies that this is equivalent to $y = z$. QED. Almost identical reasoning shows that the identity element is unique: if $x * e = x * e' = x$ for every x in G, then $e = e'$.

mathematical objects that he is studying at the moment. When he thinks he has found a proof, he starts carefully checking all his conclusions, and he realizes that only a very small number of the properties of the objects in question played any sort of role in the proof. He thus discovers that he could use the same proof for other objects having only the properties he needed to use. This demonstrates the simple underlying idea of the axiomatic method: instead of announcing which objects must be considered, it suffices to provide a list of properties [...] to be used in the investigation. Then one expresses these properties as axioms. From then on, it's no longer important to explain what objects are being studied. Instead, one can construct the proof so that it is true for any objects satisfying the axioms. It's quite remarkable that the systematic implementation of such a simple idea shook mathematics so thoroughly."

For Bourbaki, the axiomatic method is inseparable from the study of structures, the third key element of the group's vision of mathematics. What does Bourbaki mean by mathematical structures? As *L'architecture des mathématiques* explains, one starts with a set "of elements whose properties are not specified. Then one or several relations among these elements are added [...], and postulates are added that the given relation or relations must satisfy. These are the axioms of the structure in mind. Making an axiomatic theory out of a given structure is deducing logical consequences of the structure's axioms without using any other assumptions about the elements in question (and, in particular, without using any assumptions about their 'nature')."

Three Important Types of Structures

Let's illustrate the notion of structures with one of the most widespread and important mathematical structures: the group. The axioms that define groups are given in box 1. There are many concrete realizations of this abstract entity. Here are three, suggested by *L'architecture des mathématiques*:

1. The set of real numbers with the operation of ordinary addition.
2. The set of integers 1, 2, 3, 4, 5, 6 equipped with multiplication modulo 7. (The result of multiplication of two integers *m* and *n* modulo 7 is the remainder upon dividing the product *mn* by 7. For example, the result of 4 times 5 modulo 7 is 6, because 6 is the remainder when 20 is divided by 7.)
3. The set of translations of the Euclidean plane.

At the Chançay conference in 1936: Claude Chevalley (from behind), Szolem Mandelbrojt, Jean Delsarte, Jean Dieudonné, and André Weil (standing).

2. What Is a Ring? What Is an Ideal?

A ring is a set whose elements can be added, subtracted, and multiplied (but not necessarily divided). More precisely, a ring is a non-empty subset R equipped with an addition law $+$ and a multiplication law $*$ such that

1. R is an abelian group with the operation $+$. (The identity element for addition is usually denoted by 0.)
2. The law $*$ is associative and has an identity element (usually denoted by 1); that is, $x * (y * z) = (x * y) * z$ and $1 * x = x * 1 = x$ for any x, y, and z in R.
3. The law $*$ distributes over $+$; that is, $x * (y + z) = x * y + x * z$ and $(x + y) * z = x * z + y * z$ for any x, y, and z in R.

The ring R is called commutative when both the addition and multiplication laws are commutative. A familiar example of a commutative ring is the set $\mathbf{Z}$ of integers, equipped with the usual operations $+$ and $*$ of addition and multiplication. Another example of a commutative ring is the set of polynomials in one variable with coefficients in the real numbers (these are expressions of the form $a_0 + a_1x + a_2x^2 + \cdots + a_nx^n$, where n is a positive integer called the degree of the polynomial and $a_1, a_2, a_3, \ldots, a_n$ are given real numbers called the coefficients), again equipped with the usual laws of polynomial addition and multiplication. An example of a non-commutative ring is the set of $2x2$ matrices with real entries (these are tables of four numbers arranged in two lines of two elements each), which have many applications, including the representation of linear transformations acting on vector spaces. In this case, the addition and multiplication laws are the usual laws of addition and multiplication for matrices; see the diagram below.

Given a ring R, now consider a subset I of R. This subset I is a left ideal if the following properties are satisfied:

1. I is a subgroup of R with respect to the addition law $+$.
2. For any a in R and x in I, the product $a * x$ is an element of I.

The definition of a right ideal is analogous; just replace $a * x$ by $x * a$ in property 2. For example, the even numbers form an ideal (both left and right) of the ring $\mathbf{Z}$ of integers, because any multiple of an even number is even.

$$A = \begin{pmatrix} a_1 & a_2 \\ a_3 & a_4 \end{pmatrix} \qquad B = \begin{pmatrix} b_1 & b_2 \\ b_3 & b_4 \end{pmatrix} \qquad A + B = \begin{pmatrix} a_1+b_1 & a_2+b_2 \\ a_3+b_3 & a_4+b_4 \end{pmatrix} = B + A$$

$$AB = \begin{pmatrix} a_1b_1 + a_2b_3 & a_1b_2 + a_2b_4 \\ a_3b_1 + a_4b_3 & a_3b_2 + a_4b_4 \end{pmatrix} \qquad BA = \begin{pmatrix} b_1a_1 + b_2a_3 & b_1a_2 + b_2a_4 \\ b_3a_1 + b_4a_3 & b_3a_2 + b_4a_4 \end{pmatrix} \neq AB$$

With a little high school math, it is not too hard to check that these three structures indeed satisfy the group axioms. For the first example, this is nearly obvious: any real x, y, and z satisfy $(x+y)+z = x+(y+z)$, the identity element is 0 since $x+0 = 0+x = x$, and the additive inverse of x is $-x$ since $x+(-x) = 0$. For the second example, the identity element is the integer 1 and the inverses of 1, 2, 3, 4, 5, and 6 are respectively 1, 4, 5, 2, 3, and 6. For the third example, the identity element is the identity translation, which fixes every point in the plane, and each translation has an inverse that reverses the effect of that translation on the points of the plane.

Bourbaki differentiates three main types of structures. The structures involving a rule that associates any pair of elements to a third element are called algebraic structures. These sort of structures include groups, of course, as well as rings, ideals, fields, vector spaces, and so forth. The second type of structure involves an order (see box 4), which is a relation that orders, or compares, the elements of a set (although not necessarily all the elements). The structures of the third type are topological structures, which provide an abstract mathematical formulation of the intuitive concepts of *neighborhoods*, *limits*, and *continuity*. The next chapter discusses some structures of this type.

With the axiomatic method and these three main types of structures—algebraic structures, ordered structures, and topological structures—as guides, Bourbaki paints a picture of a mathematical universe where the organizing principle is a hierarchy of structures progressing from simple to complex and from general to specific. Bourbaki explains that "in the nucleus are the main types of structures [...], the mother-structures, one could say. Each of these types already contains quite a wide variety of objects, since one must differentiate the most general structure, the one with the least axioms, from the ones obtained by adding axioms, which each yield a crop of new results [...]. Beyond this innermost nucleus are the structures that one could describe as complex. These involve one or more mother-structures that aren't merely juxtaposed [...] but *combined* organically by one or more axioms connecting them [...]. Finally, farther out are the actual theories, where the elements of the sets in question, until now left undefined within the general structures, gain individual characteristics. This is where the classical mathematical theories are found." According to Bourbaki, individual theories of mathematics such as function theory, differential geometry, algebraic geometry, and number theory have lost their old autonomy to become "crossroads where the most general mathematical structures come together to act upon one another."

Having presented this vision of mathematics, the author of *L'architecture des mathématiques* emphasizes that it is merely a schematic

Pierre Cartier and Bernard Teissier in la Messuguière (near Grasse) in July 1975.

Charles Ehresmann (1905–1979)

and idealized view, and thus that it "must be treated only as a very rough approximation of the current state of mathematics."

The unity of mathematics, the axiomatic method, and the study of structures are not Bourbaki's own inventions. Mathematicians have always wondered whether or not mathematics is a single, unified subject. For instance, the question arises in simply trying to understand the connections between algebra and geometry (for example, why can a set of consecutive real numbers be identified with the points on a line segment?). The axiomatic method, pioneered by Euclid, entered modern mathematics at the end of the nineteenth century (with Hilbert's work, for example, and Richard Dedekind and Giuseppe Peano's axiomatization of integer arithmetic). As for the study of structures, at least of algebraic structures, Bourbaki was strongly influenced by van der Waerden's *Moderne Algebra*, which represented German mathematics during the years 1900–1930.

Bourbaki's role in these issues was to emphasize these three notions, to connect them, to try to extend the concept of structure emerging in the work of the German algebraists to all of mathematics. However, the Bourbakian vision of mathematics is not a perfectly constructed and coherent theory. Moreover, Bourbaki does not always follow its own philosophy of mathematics in *Éléments de mathématique.*

A working seminar at the Institut des Hautes Études Scientifiques. Grothendieck is presenting, and the first row contains (from left to right) Michel Demazure, François Bruhat, Pierre Samuel, and Jean Dieudonné (near the window).

3. What Is a Field?

A field is a set whose elements can be added, subtracted, multiplied, or divided (expecting division by zero, which is not defined). That is, a set K equipped with an addition law + and a multiplication law $*$ is a field if

1. K is a ring under + and $*$ and contains at least two elements.
2. Any nonzero element x of K has a multiplicative inverse x^{-1} (that is, x^{-1} satisfies $x * x^{-1} = x^{-1} * x = 1$). The zero element, which has no inverse, is by definition the identity element for addition.

Examples of fields are the set $\mathbf{Q}$ of rational numbers (numbers of the form p/q, where p and q are integers); the set $\mathbf{C}$ of complex numbers (the numbers of the form $x + iy$, where x and y are real and i satisfies $i^2 = -1$); and the subfield of $\mathbf{R}$ containing real numbers of the form $a + b\sqrt{3}$, where a and b are rational numbers.

Bourbaki the Ostrich

One of the main illustrations of this weakness is Bourbaki's attitude towards the axiomatization of set theory and, more globally, towards questions about the foundations of mathematics. Obtaining a satisfactory system of axioms for set theory, upon which the turn-of-the-century mathematicians wished to build all of mathematics, proved to be an arduous task that led logicians and mathematicians to much research on the foundations of mathematics. Among other things, they tried to prove that the mathematics resulting from these axioms was coherent, that the axioms could never lead to any sort of contradiction. This program, for a large part initiated by Hilbert, led to results that were surprising and hard to swallow. This is particularly the case of Kurt Gödel's proof that it is impossible to prove that a given system of axioms gives rise to a coherent system of mathematics by using only those axioms.

Faced with a crisis that was affecting mathematicians in the first third of the twentieth century, Bourbaki decided to hide its head in the sand and treat these metamathematical problems, so important to logicians, as uninteresting to mathematicians. However, it is hard to believe that the logical coherence of an axiomatic theory would be unimportant to a mathematician who, like Bourbaki, believed so much in the axiomatic method. Bourbaki's somewhat schizophrenic attitude—which is held by most mathematicians not working directly on the foundations of their discipline—materializes in the group's book

Saunders Mac Lane (1909–2005) was one of the founders of category theory.

4. What Is an Order?

The concept of an order generalizes common tools for comparison like "greater than or equal to" and "less than or equal to." More precisely, an order is a relation $\mathcal{R}$ on a set E that satisfies the following three properties.

1. Reflexivity: For any x in E, $x\mathcal{R}x$.
2. Transitivity: For any x, y, and z in E, if $x\mathcal{R}y$ and $y\mathcal{R}z$, then $x\mathcal{R}z$.
3. Antisymmetry: For any x and y in E, if $x\mathcal{R}y$ and $y\mathcal{R}x$, then $x = y$.

The most well-known example of an order is the relation "greater than or equal to" on the set of real numbers. To check this, it suffices to check that the three order axioms are satisfied when $\mathcal{R}$ is replaced by the symbol $\geq$. Another example of an order is the relation of inclusion defined on the subsets of a given set E. We say that a set A is included in B if all the elements of A are also elements of B; this is written as $A \subset B$. It is simple to check that the relation $\subset$ satisfies the three order axioms; for example, if $A \subset B$

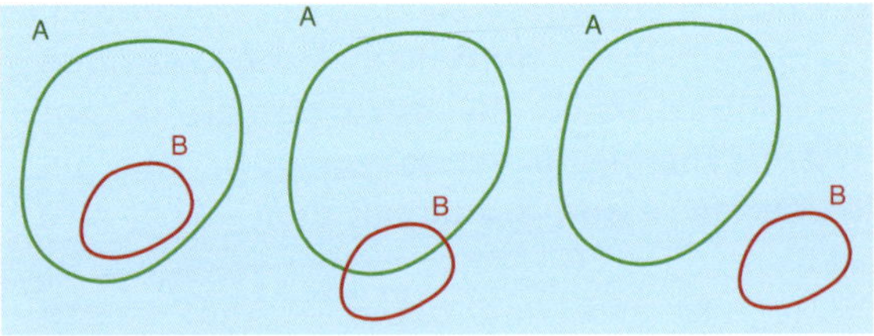

and $B \subset A$, then A and B must be the same set. Unlike the first example, where any two real numbers can be compared by the relation $\geq$ (such an order is called a total order), two sets A and B cannot necessarily be compared by the relation of inclusion. If A and B are disjoint or share some elements but not others, then neither $A \subset B$ nor $B \subset A$ is true (such an inclusion is called a partial order). A third example of an order is the relation of division defined on the set of integers (an integer a divides the integer b if b/a is an integer). This is also a partial order.

on set theory in *Éléments de mathématique.* This book met severe criticism, especially from logicians, for its narrow perspective and failure to address fundamental questions (see chapter 9).

Categories vs. Bourbakian Structures

Although Bourbaki's treatise contains numerous examples of structures, it leaves the concept of structures somewhat vague. Van de Waerden's book *Moderne Algebra,* which presented algebra as a hierarchy of structures, hadn't formalized this concept. "Van der Waerden didn't find it necessary to give an explicit explanation, be it formal or not, of what 'algebraic structures' and the 'structural study of algebra' really are," Leo Corry, a historian at the University of Tel Aviv, explains in his book. Bourbaki, on the other hand, does provide a formalization of the concept of structures in the fourth chapter of the book on set theory. However, the following chapters of the treatise do not make use of this formalization—despite the fact that the treatise's first part, consisting of the first six books, was initially called "Fundamental structures of analysis." Corry's analysis of the concept of algebraic structure as it developed since the nineteenth century "suggests that [Bourbaki's] *Théorie des ensembles* [*Theory of Sets*], and in particular the concept of *structure* it defines, are not essential to the rest of *Éléments*. It is possible to read and understand every book of Bourbaki's treatise without having learned about the concept of *structures*. In principle, *Théorie des ensembles* could have been left out of the treatise, since it has neither heuristic value nor logical importance for any of the theories discussed in the treatise's later volumes, where the real importance of the treatise lies." Also, while the treatise's goal is to provide basic tools for mathematicians, "the concept of *structures* seems forced and unnatural." Pierre Cartier, a former member of Bourbaki, agrees with this view, affirming in a study of structuralism in mathematics that "our conclusion is final: Bourbaki did not produce a mathematical theory of structures, and perhaps did not want to do so."

Jean-Louis Koszul (born 1921) during a Bourbaki conference.

Today, no discussion of mathematical structures is complete without a discussion of category theory. Introduced around 1942 by Samuel Eilenberg (who would later become a member of Bourbaki) and Saunders MacLane, category theory provides an abstract and general framework for describing numerous mathematical situations and the connections between them. Things become technical and abstract very quickly, but we can give a rough description of what category theory entails. A category is given by a class of objects and, for every pair (A, B) of objects, a set of correspondances (called morphisms) from A to B. For example, in the category of sets the objects are sets and the morphisms from A to B are all the possible functions defined from A to B. In the category of groups, the objects are groups and the morphisms from a group A to a group B are all the homomorphisms from A to B. (A homomorphism f is a function preserving the structure of the group operation; that is, a function f is a homomorphism if for all x and y in A, $f(x * y) = f(x) \circ f(y)$, where $*$ is the composition law of the group A and $\circ$ is the composition law of the group B.) In addition, category theory studies the mapping between two categories, which are called functors.

The language of categories and functors spread rapidly during the sixties. Some Bourbaki members put it to great use, including Eilenberg of course, but also Charles Ehresmann and especially Alexandre Grothendieck. Category theory, which is much more general than the structures described by Bourbaki in *Éléments de mathématique*, could have played an important part in the structural vision of mathematics, but Bourbaki did not update its *Architecture des mathématiques.* More importantly, the group did not manage to use categories in its treatise, despite the group's numerous discussions and preliminary drafts on the subject. One of the reasons for this is that the task would have required a profound revision of the existing volumes. According to Pierre Cartier, "Bourbaki got away with talking about categories without really talking about them. If they were to redo the treatise, they would have to start with category theory. But there are still unresolved problems about reconciling category theory and set theory."

In her thesis on Bourbaki, Judith Friedman collected these remarks from Claude Chevalley, one of the group's founders, which relate to the idea that Bourbaki has become mathematically more bourgeois over the years: "From this point of view, category theory was more true to the spirit of Bourbaki than their theory based on structures—it was more structuralist! [...] In some sense, Bourbaki's rejection of categories was one of the most significant points in the transformation of the group's spirit. For the first time, something that people knew to be eminently Bourbakian was mostly rejected out of a desire to advance without addressing [...] the starting point." To try to rectify this, Chevalley wrote a book on category theory in the 1960s. However, the book was never published—the publishers, Hermann, lost the manuscript!

The fifties and sixties were the years of structuralism. The Bourbakian vision, which emphasized axioms and structures, was emulated by mathematicians as well as authors (particularly the Oulipo movement) and scientists in areas such as anthropology and psychology. Many criticized Bourbaki's vision of the structure of mathematics, and it is clear that their vision cannot encompass all the mathematical activity of today (see chapter 11). Nevertheless, this vision certainly encompassed the mathematics of their time and brought some clarity to the era in which it was conceived and expressed.

The logician Kurt Gödel and the physicist Albert Einstein at Princeton in 1950.

6

A Snapshot of Bourbaki's Work: Filters

Nicolas Bourbaki prided itself in scrutinizing current mathematics rather than in inventing new mathematics. Nonetheless, Bourbaki's cauldron did produce some new concepts, including a mathematical structure called a filter.

In deciding to write a treatise on analysis, Bourbaki had given itself a mathematical task similar to "cleaning the stables of Augias," as André Weil described it. This task involved scrutinizing and clarifying basic concepts, scrupulously developing terminology, reorganizing topics, and choosing a certain point of view every time there were several possible choices (which is often the case in mathematics). On the other hand, the task did not involve inventing new mathematics, and, strictly speaking, Bourbaki's treatise did not introduce new discoveries. Yet this statement must be qualified. Of course, each of the Bourbakis individually contributed new discoveries towards the development of mathematics. Furthermore, the collective work carried out by Bourbaki inspired some new mathematical concepts. In this chapter, we will discuss an example of such an innovation that does not require too much techical knowledge: the concept of a filter.

Filters were invented by Henri Cartan in September 1937, during a Bourbaki conference held in Chançay (near Tours) at the home of Claude Chevalley's parents. One day, the mathematicians attending this meeting were working on certain problems in topology, in particular searching for a way to free the definition of convergence from using the concept of countability. As Liliane Beaulieu writes in a 1990 article, the group decided to take a break and go for a walk at a point when the discussion seemed to be getting nowhere. Henri Cartan, however, decided to stay back and continue thinking. This led Cartan to a new idea, which he presented to his friends upon their return. At first they met the idea with scepticism, but Chevalley understood the importance of Cartan's suggestion and even proposed another idea based on it (which became the concept of ultrafilters). Once the approval was unanimous, someone yelled *boum*! (French for "bang!") to announce that a breakthrough had been made—this was one of Bourbaki's many customs. Thus Cartan's invention became known as *boum*, and the theory surrounding it became *boumologie*. Cartan reported the discoveries made during the Chançay conference in two articles sent a few weeks later to the *Comptes Rendus de l'Académie des Sciences*. As members of the Académie ought to be shown proper respect, this was probably one of the reasons that *boums* were rebaptised as *filtres*, which became "filters" in English.

Of course, Cartan's invention was included in Bourbaki's book on general topology. This and other innovations (in particular, André Weil's uniform structures and Jean Dieudonné's paracompact spaces) contributed to the success earned by this book of *Éléments de mathématique*, which was one of the most unanimously appreciated Bourbaki volumes at the time.

At the 1937 Bourbaki conference in Chançay. From left to right are André Weil, Simone Weil (standing), Henri Cartan, Szolem Mandelbrojt, and Claude Chevalley. Jean Delsarte's head is just barely visible at the far right.

So what are filters? Before discussing what these structures are, we must discuss a few preliminary concepts of topology and analysis. One common description of topology is that it studies everything about neighborhoods, limits, and continuity. Another is that it studies the properties of geometric or abstract mathematical objects that remain unchanged when the object is deformed continuously, without pinching or tearing it. Either way, the concept of continuity (or, more precisely, that of a continuous function) is the cornerstone of this branch of mathematics. A rigorous definition of continuity relies on a rigorous definition of limits. Filters come into play with the definition of limits, and they lead to a very abstract and general formulation of this definition.

Let's start with an explanation of continuous functions, which are fundamental objects in mathematical analysis. Consider a real-valued function defined on the set **R** of real numbers; in other words, consider a function f such that x and $f(x)$ are both real numbers. Everyone has learned to draw a graph of this sort of function on an x, y-coordinate plane in school: the graph of the function f is the set of points with coordinates (x, y) such that $y = f(x)$. So when is such a function continuous? Intuitively, a function is continuous if one can trace its graph without ever lifting the pencil from the paper (see figure 1). This is true for most functions you might encounter, like $f(x) = x^2$ or $f(x) = \sin x$. Meanwhile, the function defined by $h(x) = 0$ whenever $x \leq 0$ and $h(x) = 1$ whenever $x > 0$ is not continuous (see figure 2). In particular, this function has a discontinuity at $x = 0$, since the graph "jumps" at this point. There are even functions that have so many discontinuities that they are impossible to draw. This is case for the function taking the value 0 whenever x is rational and 1 whenever x is irrational: it has a discontinuity at every point! Another strange example is the function f defined by $f(x) = 0$ whenever x is irrational, $f(x) = 1/q$ whenever $x = p/q$ (with p and q being relatively prime) is a nonzero rational number, and $f(0) = 1$. This function is continuous at every irrational point and discontinuous at every rational one!

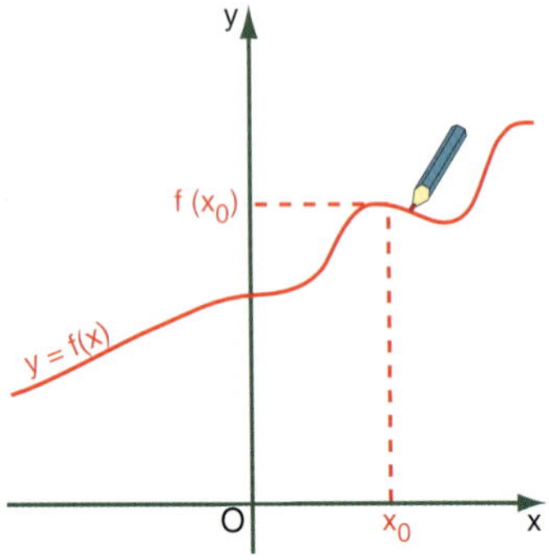

1. The graph of a continuous function, $y = f(x)$

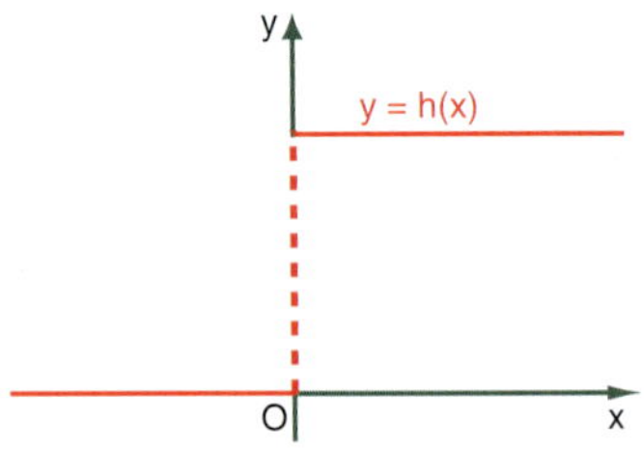

The function h is not continuous at $x = 0$.

Of course, the definition of continuity given above is too vague for a mathematician. Here is a more satisfactory definition. The function f is continuous at a given point x_0 if $f(x)$ approaches $f(x_0)$ as x approaches x_0. This is equivalent to saying that the limit of $f(x)$ as x approaches x_0 is $f(x_0)$, which is written as $\lim_{x \to x_0} f(x) = f(x_0)$. Roughly, this means that the closer x is to x_0, the closer $f(x)$ is to $f(x_0)$, and that $f(x)$ gets infinitely close to $f(x_0)$. It is easy to see that in figure 1 this property is satisfied at every point x_0 and that in figure 2 this property is not satisfied at the point $x_0 = 0$.

1. The Distance Axioms

Let E be a set. A distance (also called a metric) on E is a function d that associates to each pair (X, Y) of points in E a nonnegative number such that the following axioms are satisfied for all X, Y, and Z in E:

1) Separation: $d(X, Y) = 0$ if and only if $X = Y$.
2) Symmetry: $d(X, Y) = d(Y, X)$.
3) Triangle inequality: $d(X, Y) \leq d(X, Z) + d(Z, Y)$.

A set equipped with a distance function is called a metric space. For example, the set $\mathbf{R}$ of real numbers forms a metric space when equipped with the distance function $d(x, y) = |x - y|$. The set $\mathbf{R}^2$ of pairs of real numbers is also a metric space, this time when equipped with either the standard Euclidean distance or the function $d(X, Y) = |x_1 - y_1| + |x_2 - y_2|$, where $X = (x_1, x_2)$ and $Y = (y_1, y_2)$.

The definition of continuity using limits would be perfectly rigorous and operational as soon as the concept of limits was rigorously defined. However, what we have said about limits is not terribly precise. For example, what exactly do the phrases "x approaches x_0" and "$f(x)$ gets infinitely close to $f(x_0)$" mean? We will explain this using an un-Bourbakian method, by progressing from specific to general instead of vice-versa, as Bourbaki would usually advise.

Let's begin by discussing the limit of an infinite sequence $u_0, u_1, u_2, \dots, u_n, \dots$ of real numbers. To simplify the notation, we write such a sequence as $(u_n)_{n\geq 0}$, with u_n being called the general term of the sequence. For example, the sequence $1, 1/2, 1/3, 1/4, 1/5, \dots, 1/n, \dots$ is written simply as $(1/n)_{n\geq 1}$, and the sequence $0, 1, 4, 9, 16, 25, \dots, n^2, \dots$ is written as $(n^2)_{n\geq 0}$. Now, it is clear that $1/n$ decreases as n increases, and that $1/n$ approaches 0 in "the limit" where n approaches infinity. Similarly, it is easy to notice that the sequence $1, 3/2, 5/3, 7/4, \dots$ with general term $(2n+1)/(n+1)$ approaches 2 as n approaches infinity. On the other hand, the sequence $(n^2)_{n\geq 0}$ does not approach a finite number as n approaches infinity; in this case, we say that the sequence has no limit, or diverges. While these assertions about limits are intuitively obvious, they require rigorous formulations. So here is the rigorous definition of limit for the case of infinite sequences:

A sequence $(u_n)_{n\geq 0}$ of real numbers approaches (or converges to) a limit L as n approaches infinity if for any positive ϵ there exists a positive integer n_0 (which may depend on ϵ) such that $|u_n - L| < \epsilon$ for every $n \geq n_0$. If $(u_n)_{n\geq 0}$ approaches L, we write that $\lim_{n\to\infty} u_n = L$.

Such definitions are always hard for newcomers to understand; however, the idea is really rather simple. Basically, it says that no matter how small ϵ is, there always exists an n_0 such that all terms u_n with

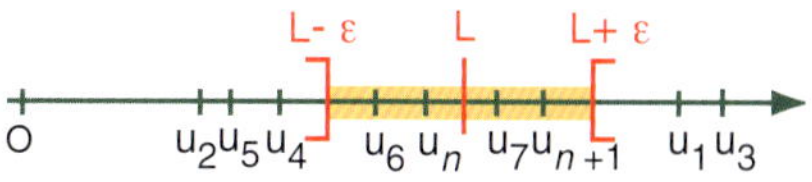

3. The sequence u_n approaches L if for every positive ϵ there exists an integer n_0 such that $|u_n - L| < \epsilon$ whenever $n \geq n_0$.

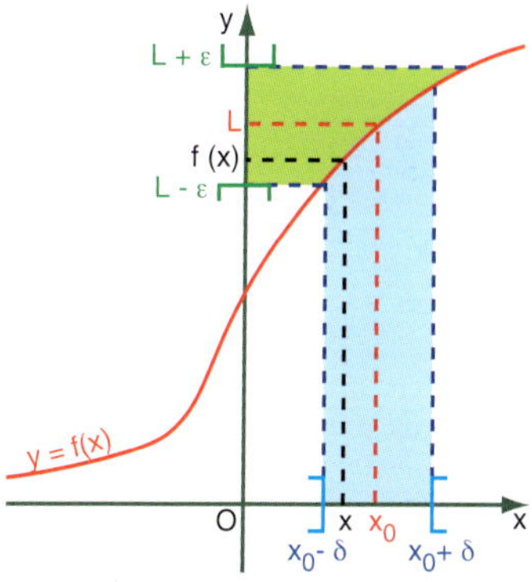

4. The limit of the function f as x approaches x_0 is L if for any $\epsilon > 0$ there exists a $\delta > 0$ such that $|f(x) - L| < \epsilon$ whenever $|x - x_0| < \delta$.

2. Topological Spaces, Open Sets, and Neighborhoods

A topological space is a set E with a finite or infinite collection of subsets of E (called open sets) such that the following axioms are satisfied:

1) E and the empty set $\varnothing$ are open sets.
2) The intersection of any two open sets is an open set.
3) The union of any number of open sets is an open set.

These axioms define a structure called a topology on the set E. The elements of the topological space are called points, in analogy with the points of spaces in geometry.

Given a topological space E, a subset V of E is called a neighborhood of the point a if V contains an open set containing the point a.

The concepts of topological spaces, open sets, and neighborhoods are of fundamental importance to the subject.

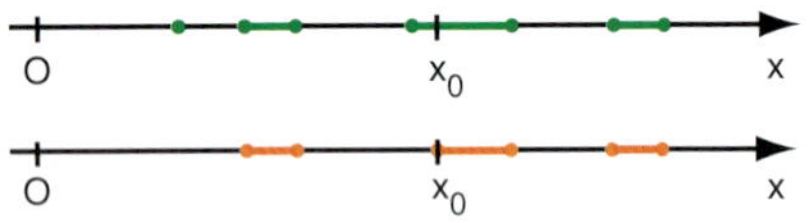

5. The subset of R colored in green is a neighborhood of x_0, while the one colored in red is not.

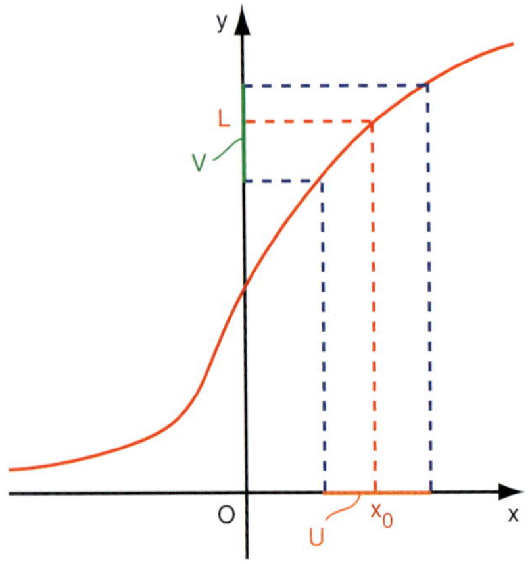

6. For any neighborhood V of L, there is a neighborhood U of x_0 such that $f(x) \in V$ for any $x \in U$.

$n \geq n_0$ are no more than ϵ away from L (see figure 3, p. 89). This is equivalent to saying that all terms u_n with $n \geq n_0$ lie in the open interval $(L - \epsilon, L + \epsilon)$. Let's illustrate this with the sequence $(1/n)_{n\geq 1}$. For $\epsilon = 0.1$, we can set n_0 to be 15, since the distances between 0 (the hypothesized limit of the sequence) and $1/15$, $1/16$, $1/17$, and so on are all less than 0.1. Similarly, for $\epsilon = 0.01$, we can set n_0 to be any integer greater than 100. In fact, for any epsilon, setting n_0 to be any integer greater than $1/\epsilon$ guarantees than $1/n$ is at most ϵ away from 0 for every $n \geq n_0$. This proves that 0 is indeed the limit of the sequence.

It is simple to extend the definition of limit for sequences of real numbers to a definition of limit for functions $f : \mathbf{R} \to \mathbf{R}$. This generalized definition is:

The function f has limit L as x approaches x_0 if for any positive ϵ there exists a positive δ (which may depend on ϵ) such that $|f(x) - L| < \epsilon$ for every $x \neq x_0$ satisfying $|x - x_0| < \delta$.

In other words, $f(x)$ can be made as close to L as one wants (within ϵ of L) as long as x is close enough to x_0 (within δ of x_0). Equivalently, $f(x)$ lies in the open interval $(L - \epsilon, L + \epsilon)$ whenever x lies in the open interval $(x_0 - \delta, x_0 + \delta)$ (see figure 4, p. 89).

This definition of limit allows us to formulate a rigorous definition of continuity:

A function $f : \mathbf{R} \to \mathbf{R}$ is continuous at x_0 if f has a limit as x approaches x_0 and this limit equals $f(x_0)$, the value of f at x_0.

To help generalize the concept of limits even more, it is useful to reformulate the above definition of limit for functions in terms of neighborhoods. First, what is a neighborhood? If x_0 is a real number, we say that a subset V of $\mathbf{R}$ is a neighborhood of x_0 if this subset V contains an open interval containing x_0 (see figure 5). Recall that an open

3. The Filter Axioms

Let E be a set. A nonempty set F of subsets of E is called a filter on E if it satisfies the following three axioms:

1) F does not contain the empty set.
2) If A is an element of F that is contained in a subset B of E, then B is also an element of F.
3) If A and B are elements of F, then their intersection $A \cap B$ is also an element of F.

Here is an example of a filter. Let E be a topological space and let F be the set of neighborhoods of a given point x. It is easy to check that

1) The empty set is not a neighborhood of x,
2) If A is a neighborhood of x, then every set B containing A is also a neighborhood of x, and
3) If A and B are neighborhoods of x, then $A \cap B$ is also a neighborhood of x.

Thus the set of neighborhoods of x is a filter on E.

interval is an interval (a, b), which is the set of all real x satisfying $a < x < b$. By examining the definition of limit for a function $f : \mathbf{R} \to \mathbf{R}$, we see that we can restate it as:

The function f has limit L as x approaches x_0 if for any neighborhood V of L there exists a neighborhood U of x_0 such that $f(x)$ lies in V for any $x \neq x_0$ lying in U (see figure 6).

The advantage of this formulation is that it doesn't only apply to functions defined from $\mathbf{R}$ to $\mathbf{R}$. Until now, we have only considered functions that associate $f(x)$ to a real number x. Yet in mathematics and its applications, it is often useful to consider functions $f : X \to Y$, where X and Y are sets other than $\mathbf{R}$. For example, suppose one wanted to represent the movement of an object in a plane as a function of time (see figure 7). At any given moment, the position of the object is represented by a point with coordinates (x, y). Since these coordinates change with time, the position of the object at time t is represented by $(x(t), y(t))$, where $x(t)$ and $y(t)$ are functions taking values in $\mathbf{R}$. The function f associating to each time t the position $(x(t), y(t))$ of the object at that time is a function defined on $\mathbf{R}$ and taking values in $\mathbf{R}^2$ (the set of pairs of real numbers). In symbols, this is written as $f : \mathbf{R} \to \mathbf{R}^2$. Similarly, one can think of functions from $\mathbf{R}^2$ to $\mathbf{R}$ (which associate some real number to each pair of real numbers), functions from $\mathbf{R}^3$ to $\mathbf{R}^2$ (which associate some pair (y_1, y_2) of real numbers to each triple (x_1, x_2, x_3) of real numbers), and more generally functions from $\mathbf{R}^m$ to $\mathbf{R}^n$, where m and n are two given integers (these functions associate some n-tuple $(y_1, y_2, \ldots, y_n)$ of real numbers to each m-tuple $(x_1, x_2, \ldots, x_m)$ of real numbers).

Can the concepts of limits and continuity be defined for such functions? Indeed they can, simply by using the definitions given above.

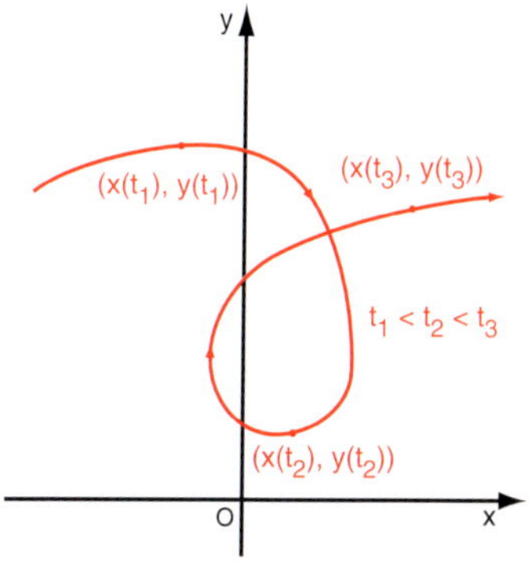

7. Here the trajectory of an object moving in a plane is represented by the function f from R to $\mathbf{R}^2$ associating each time t to the position $(x(t), y(t))$.

All we need to do is define the concept of neighborhood for spaces of the form $\mathbf{R}^n$. In $\mathbf{R}$, a neighborhood of the point a is a subset of $\mathbf{R}$ that contains an open interval containing a, so we must give the equivalent of open interval to define neighborhood in the general case.

Let's first consider the case of $\mathbf{R}^2$, which can be represented by the plane. In this case, the equivalent of an open interval is an open disk. The open disk with center C and radius $r > 0$ is, by definition, the set of points $X \in \mathbf{R}^2$ such that the distance from each point to the center C is strictly less than r; in other words, this open disk is the set of points inside the circle with center C and radius r. Letting $d(A, B)$ be the distance between two points A and B in the plane, the open disk with center C and radius r is the set of points X such that $d(X, C) < r$ (see figure 8). The distance function d can be taken to be the standard Euclidean distance defined as $d(X, C) = \sqrt{(x_1 - c_1)^2 + (x_2 - c_2)^2}$, where X is the point (x_1, x_2) and C is the point (c_1, c_2). But in fact any of a number of possible functions qualify as a distance function, since any function satisfying a certain set of axioms is a distance function (see box 1, p. 89).

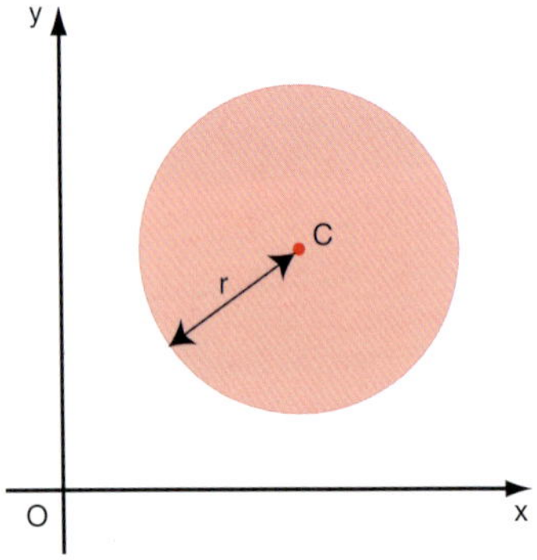

8. The open disk with center C and radius r.

Now that we have defined open disks, we can define open subsets by saying that a subset A of $\mathbf{R}^2$ is open if for every point X in A there is an open disk with center X and positive radius that is entirely contained in A. Then a subset V of $\mathbf{R}^2$ is a neighborhood of a point P in V contains an open set containing P (see figure 9). Using these definitions, the above definition of a limit for a function applies to functions $f : \mathbf{R} \to \mathbf{R}^2$ (and to functions $f : \mathbf{R}^2 \to \mathbf{R}^2$, by letting x and x_0 be points in $\mathbf{R}^2$).

It is easy to generalize these definitions for the specific case of $\mathbf{R}^2$ to definitions for the general case of $\mathbf{R}^n$. In fact, nothing in the above definitions change except for the concrete formulas for the distance $d(X, Y)$ between two points and the term "open disk," which is replaced by "open ball" in the general case. For example, in $\mathbf{R}^3$ the open ball with center C and radius r is the set of points X satisfying $d(X, C) < r$. If the distance function d is taken to be the standard Euclidean distance, this is the set of $X = (x_1, x_2, x_3)$ satisfying $\sqrt{(x_1 - c_1)^2 + (x_2 - c_2)^2 + (x_3 - c_3)^2} < r$, where $C = (c_1, c_2, c_3)$.

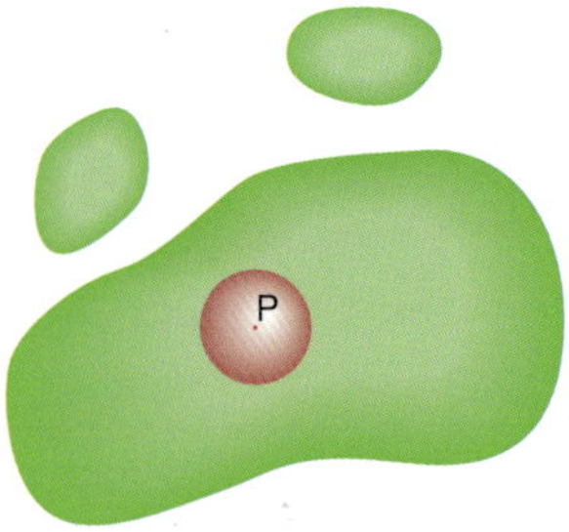

9. The green region is a neighborhood of P since it contains an open disk with center P and nonzero radius.

In fact, the above definitions can be generalized to any set with a distance function, since the definition of open sets and neighborhoods depends only on the existence of a distance function defined on the set. Sets equipped with a distance function are called metric spaces, and the spaces $\mathbf{R}^n$ are just a few examples of such spaces. For the sake of generality, here are the definitions of limit and continuity for functions $f : X \to Y$, where X and Y are any metric spaces:

The function $f : X \to Y$ (where X and Y are metric spaces) has limit

$L(\in Y)$ as $x(\in X)$ approaches $x_0(\in X)$ if for any neighborhood of L there exists a neighborhood U of x_0 such that $f(x)$ is an element of V for every $x \neq x_0$ lying in U (see figure 10).

A function $f : X \to Y$ is continuous at x_0 if f has a limit as x approaches x_0 and this limit equals $f(x_0)$, the value of f at x_0.

Of course, in these definitions the symbols x, x_0, L, $f(x)$, and $f(x_0)$ are not necessarily usual numbers: x and x_0 are elements of the space X, while L, $f(x)$, and $f(x_0)$ are elements of the space Y.

Yet our pursuit of abstraction and generality does not end here. Sometimes mathematicians need to work with sets that are not metric spaces, with sets on which no distance function is defined. Such cases led mathematicians to generalize the concepts discussed above to an even greater extent, which resulted in completely axiomatic definitions of open sets and neighborhoods independent of any concept of distance. The sets considered in these definitions are called topological spaces and include metric spaces as a special case. In some sense, this axiomatic definition extracts the essence of the properties of the open sets and neighborhoods appearing in metric spaces (see box 2, p. 90).

We have almost reached the goal of our efforts, and we are almost ready to attack filters. When X and Y are topological spaces, the definitions of limits and continuity are exactly the same as those given above for the case of metric spaces. However, some complications appear when the function f is considered on a subset A of X rather than on all of X. For example, it is common to consider functions defined only on a finite interval of **R**. Consider the function $f(x) = \sqrt{1 - x^2}$, which is only defined on the interval $[-1, 1]$ (since in the real numbers $1 - x^2$ must be nonnegative to have a real square root). Now, to study the limits and continuity at the extremities of the interval—that is, at the points $x = -1$ and $x = 1$—it would, in principle, be necessary to consider the neighborhoods of these points. But these neighborhoods cannot extend out of the set on which the function is defined, so in this case they cannot extend left of $x = -1$ or right of $x = 1$. To ensure this, we must consider only the part of each neighborhood that is contained in the interval of definition. In other words, a neighborhood U of (for example) $x = 1$ must be restricted to $U' = U \bigcap [-1, 1]$ (the intersection of U with the interval) to make sure that it doesn't extend outside the interval. In the general case of a function defined on a subset A of a topological space X, one must consider the sets $U \bigcap A$ rather than the neighborhoods U, which could extend outside of A. These considerations lead to a modified definition of limit:

Let X and Y be topological spaces, A a subset of X, and f a function mapping from A to Y. The function f has limit $L(\in Y)$ as $x(\in A)$

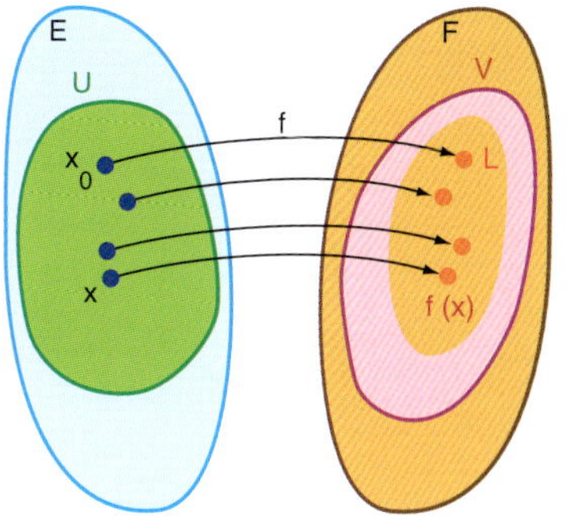

10. The limit as x approaches x_0 of a function f from E to F is L if for any neighborhood V of L there exists a neighborhood U of x_0 such that $f(x) \in V$ for any $x \in U$.

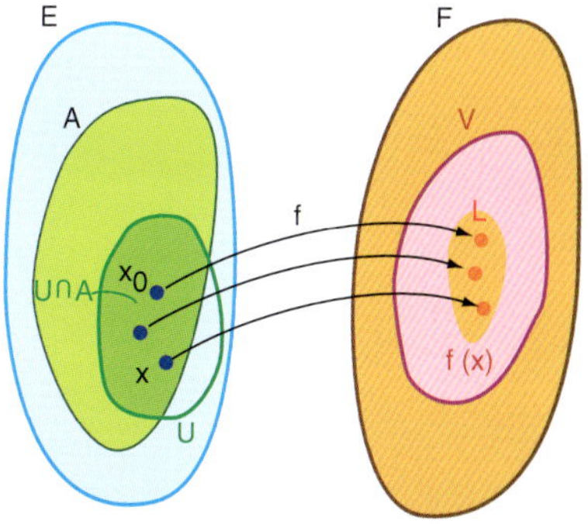

11. The limit as x approaches x_0 of a function f from A to F is L if for any neighborhood V of L there exists a neighborhood U of x_0 such that $f(x) \in V$ for any $x \in U \cap A$.

approaches $x_0(\in A)$ if for any neighborhood V of L there exists a neighborhood U of x_0 such that $f(x)$ is in V for every $x \neq x_0$ lying in $U \bigcap A$ (see figure 11, p. 93).

The general properties satisfied by sets of the form $U \bigcap A$ lead to the abstract and axiomatic definition of structures called filter bases:

Let A be any set. A nonempty family B of subsets of A is called a filter base on A if

1. The empty set $\varnothing$ is not contained in B;
2. The intersection of any two elements of B contains an element of B.

It is simple to check that if A is a subset of a topological space X and a is a point in A, then the sets of the form $U \bigcap A$ (with U being a neighborhood of a in X) form a filter base on A. This suggests a generalization of the concept of limits formulated in terms of filter bases:

Let A be any set, B a filter base on A, and Y a topological space. The limit of a function $f : A \to Y$ with respect to the filter base B is $L(\in Y)$ if for any neighborhood V of L there exists a set S in the filter base B such that $f(x)$ lies in V for every x in S.

This is a difficult and abstract definition. One important thing to understand about it is that all our previous definitions of limit are special cases of this definition. In addition, note that the set A doesn't even need to be a topological space. Finally, we referred to filter bases rather than filters to simplify the discussion; see box 3, p. 91 for the definition of filters.

And what is the point of filters? As we discussed above, these objects allow the construction of a vastly generalized sort of limit, one so general that it applies to functions defined on almost any set. This allows theorems formulated in terms of standard limits (for example, limits of sequences) to apply under less restrictive conditions when formulated in terms of filters. For example, there is a theorem asserting that a function f from a metric space X to a topological space Y is continuous at a point $a \in X$ if and only if the sequence $(f(u_n))_{n\geq 0}$ approaches $f(a)$ for every sequence $(u_n)_{n\geq 0}$ approaching a. If we rewrite this theorem in terms of filters, not only do we free ourselves from the use of countability (a concept related to sequences), but we also make the theorem true for any topological space X (instead of just for any metric space X). Thus filters allow certain topological properties to be extended from metric spaces to all topological spaces. Furthermore, filters have found applications in areas other than topology, notably, in mathematical logic.

Henri Cartan (born 1904)

Henri Cartan

Henri Cartan was born on July 8, 1904 in Nancy, where his father, Élie Cartan, was a professor of mathematics. After his father was appointed to the Sorbonne in 1909, Henri studied at the Lycée Buffon in Paris and then at the Lycée Hoche in Versailles. He was accepted into l'École Normale Supérieure in 1923. Upon graduating in 1926, Henri Cartan obtained a fellowship allowing him to work on his thesis, which he finished two years later. He was a high school teacher in Caen from 1928 until April 1929 before becoming a lecturer at the University of Strasbourg. Then, after two years at the Faculty of Sciences of Lille, he returned to Strasbourg in 1931. In 1935, he married Nicole Weiss, the daughter of the physicist Pierre Weiss, who studied magnetism. Henri Cartan moved to Clermont-Ferrand in September 1939, when the beginning of World War II caused the University of Strasbourg to be evacuated to that city. In 1940, Cartan went to Paris upon being appointed to the Sorbonne. He taught mostly at l'École Normale Supérieure until 1965, except for the year between 1945 and 1947, when he taught at Strasbourg in fulfillment of a promise he had made to this town's rector during the war. He later stopped teaching at the ENS (where he quietly but effectively reformed the teaching of mathematics) to teach at the Sorbonne. In 1969 he became a professor in the Faculty of Sciences at Orsay, where he worked until his retirement in 1975.

Most of Henri Cartan's work deals with functions of several complex variables. Most notably, he introduced sheaves, a concept invented by Jean Leray in algebraic topology, into the geometry of analytic spaces. He also studied topology (where he invented the concept of filters) and a subject called potential theory. Among his honors is the Wolf Prize, which he received in 1980. Discussing Henri Cartan and Bourbaki, Armand Borel wrote in 1998 that "for us, Henri Cartan was the most striking illustration, almost an incarnation, of Bourbaki. He was astoundingly productive despite his many administrative tasks and teaching at l'École Normale Supérieure. None of his work [...] seemed to use radically new or revolutionary ideas. Instead, through a truly Bourbakian approach, Cartan proved consecutive natural lemmas that would suddenly lead up to a great theorem. One time, I commented on Cartan's output to Serre, and he replied, 'Well, all right, it was twenty years of goofing around with Bourbaki, and there you go.' "

Henri Cartan also spent a lot of energy in fighting for an ideal he called dear to his heart: solidarity among the people of Europe. To this end, he got back in touch with German colleagues as soon as the war was over, despite the fact that his brother Louis, who had participated in the Resistance, was deported and killed by the Nazis in 1943. He also participated actively in the European Association of Teachers, which was created in Paris in 1956. As president of the French chapter in 1960, Henri Cartan organized the first meeting of mathematicians from eight European countries with the goal of finding ways to coordinate mathematics teaching and to make student exchanges possible.

Élie Cartan (1869–1951), the father of Henri Cartan, was one of the great mathematicians of the first half of the twentieth century.

7

The Bourbaki Seminar

Every year since 1948, Bourbaki has chosen speakers and subjects for talks to be given at three weekend seminars in Paris. Nearly two hundred mathematicians attend each event.

Séminaire Nicolas BOURBAKI
Institut Henri Poincaré
(Amphithéâtre Hermite), 11 rue Pierre et Marie Curie, Paris 5e.
12-13 juin 1999

PROGRAMME
Samedi 12 juin 1999
(14h30) Y.I. MANIN - Classical computing, quantum computing and Shor's factoring algorithm
(16h00) R. BRYANT - Recent advances in the theory of holonomy

Dimanche 13 juin 1999
(11h00) J. OESTERLE - Densité maximale des empilements de sphères en dimension 3 (d'après Thomas C. Hales et Samuel P. Ferguson)
(14h00) C. BREUIL - Intégration sur les variétés p-adiques (d'après Coleman et Colmez)
(15h30) J.-P. SERRE - Sous-groupes finis des groupes de Lie

(Des brochures contenant les cinq exposés de ce Séminaire seront distribuées au début de chaque séance : 500 exemplaires seront disponibles au cours de cette session)

An announcement for the June 1999 Bourbaki seminar, held at the Henri Poincaré Institute in Paris.

It was two o'clock on the Saturday afternoon of June 12, 1999. At such a time the little campus of the Henri Poincaré Institute in the Latin Quarter of Paris would usually be deserted. That day, however, it was not. Already, around twenty mathematicians had gathered in the Institute's entry hall. Most were milling around idly; they were waiting for the first lecture of the Bourbaki Seminar to begin. As this lecture was scheduled for 2:30, these mathematicians had arrived a half-hour early. Yet some had already settled in the Hermite Auditorium, located just a few meters from the Darboux Auditorium, where the founders of Bourbaki attended the Julia Seminar as young students in the 1930s (see the box on page 102).

Little by little, the crowd of mathematicians grew. There were mathematicians both young and old, both from France and from English-, German-, and Russian-speaking countries; however, there were very few women, as is so often the case in sciences like math and physics. The mathematicians reunited with old friends and struck up conversations with new acquaintances; the hubbub swelled. The French Mathematical Society and another math book publisher exhibited at stands near the entrance to the auditorium.

The Rituals of the Seminar

Around 2:20, a car parked across from the Henri Poincaré Institute. The man who climbed out was the mathematician Joseph Oesterlé, at the time the director of the Institute and currently a prominent member of Bourbaki. With some assistance, he unloaded several heavy

Yuri Manin's talk on quantum computing (2:30–3:30) at the Bourbaki seminar, given on June 12, 1999 in the Hermite Auditorium at the Institut Henri Poincaré.

boxes and carried them into the auditorium. This caused great excitement, especially when the contents—five hundred copies of the one-hundred-paged booklet containing the lectures to take place at this Bourbaki seminar—were unpacked and distributed. The crowd thronged as everyone collected a copy for himself, and often several for friends.

Just a few minutes later, most of the mathematicians were seated in the auditorium, discussing or reading the contents of the booklet they had just received. Oesterlé was seated at the right end of the first row, near the overhead projector. The audience also included many other notable figures of French mathematics, including Marcel Berger, Jean-Pierre Bourguignon, Pierre Cartier (a former Bourbaki), Adrien Douady (also a former Bourbaki), and the young Maxim Kontsevitch (a 1998 Fields medallist). At 2:30 exactly, the person standing near the blackboard started to talk in English with a Russian accent, and silence fell over the audience. And thus the first lecture started without any sort of introduction by the seminar's organizers, which is very unusual for this sort of meeting. The speaker was Yuri Manin, a Russian mathematician working at the Max Planck Institute in Bonn, Germany. Titled "Classical computing, quantum computing, and Shor's factoring algorithm," Manin's talk was devoted to quantum computing, a recent area of physics research that studies a new (but still hypothetical) type of computer based on the principles of quantum physics.

Manin's talk, given on transparencies as is common in many mathematical conferences, started at an elementary level but quickly became more technical. An hour later, Manin gave in to an American tradition of ending one's lecture with a humorous transparency. In Manin's case, the transparency showed a caricature of a man dismounting an axeblade whose interior was as complex as that of a clock. He declared that what he liked about the picture was that there wasn't any clear moral to the story, which provoked laughs and smiles followed by the traditional applause. The applause erupted again after the no less traditional question and answer session, and Manin's presentation was complete.

A short break followed, which some used to make purchases at the book publishers' stands or to slip away, while others reunited and discussed away. At four o'clock, the seminar reconvened immediately with a talk by Robert Bryant, a young American from Duke University. Like the first talk, this one was given in English, but this time at the blackboard instead of on transparencies. The auditorium was less full than during the first lecture. Was this because the speaker was less famous, the subject ("Recent advances in the theory of holonomy," an area pioneered by Élie Cartan, the father of Bourbaki) aroused less curiosity, or because the booklet containing the lecture had already been distributed?

Bryant's lecture was the last of the afternoon. The next afternoon, our mathematicians met again for three more hour-long lectures, this time all given in French. The first was Joseph Oesterlé discussing the *Densité maximale des empilements de sphères en dimension 3* [*d'après Thomas Hales et Samuel Ferguson*] ("Maximum sphere packing density in dimension 3 [according to Thomas Hales and Samuel Ferguson]"). Christophe Breuil, of the Université de Paris-Sud, followed with his talk on *Intégration sur les variétés p-adiques* [*d'après Coleman, Colmez*] ("Integration on p-adic varieties [according to Coleman, Colmez]"). The final talk of the seminar was given by the eminent Jean-Pierre Serre, a professor at the Collège de France and a longtime member of Bourbaki, who talked about *Sous-groupes finis des groupes de Lie* ("Finite subgroups of Lie groups").

The Bourbaki seminars: 864 published lectures

864 Lectures and 10,000 Pages

Published in the booklet for this June 1999 Bourbaki seminar, Serre's lecture is numbered 864. It's not surprising that there have been so many lectures, since the seminar has occurred three times a year (roughly in February, June, and November) since the first seminar in December 1948. The seminar was initially held at l'École Normale Supérieure, where Henri Cartan taught, but it was later moved to the Henri Poincaré Institute, where a larger lecture room was available. Until 1987, seminars usually consisted of six lectures spread over three half-

In front of the Henri Poincaré Institute during the break from 3:30 to 4:00.

days (on Saturday, Sunday, and Monday); however, the group was forced to abandon the Monday sessions and limit the number of lectures to five because of the scientists' ever-busier schedules.

All these seminars yield an impressive total of 864 lectures, collected into books and printed on about 10,000 pages. But what is the importance of the Bourbaki Seminar, which has become the only visible activity of the group since the last addition to *Éléments de mathématique*? The seminars don't convey the spirit of Bourbaki in the same way that the group's treatise and pranks used to, and the only element left of the old Bourbaki is the attempt to keep the members' names secret. Instead, the seminars seem to be pretty standard conferences on mathematical subjects of current interest, similar to so many mini-colloquiums of the scientific community. Considering all this, why has the Bourbaki Seminar been so succesful?

When asked about this, most mathematicians answer that the Bourbaki seminars are the only non-specialized seminars in France. In fact, these seminars are almost the only ones that present current research in a way that is accessible to mathematicians specializing in all different fields. (They are just "almost" the only ones because, for example, the French Mathematical Society has devoted the afternoon of their *Journée Annuelle*, the day each year when the General Assembly meets, to two or three lectures on mathematical subjects of general interest.) The presentations at the Bourbaki seminars are relatively accessible thanks in part to the choice of lecturers. For one thing, the lecturers are generally not specialists in the subject of their lecture, so they are

Kepler conjectured that the densest method of stacking identical spheres was the "fruit merchant" method represented below. Thomas Hale proved this conjecture in 1998, and Joseph Oesterlé presented the result at the June 1999 Bourbaki seminar.

Pierre Samuel

more aware of any difficulties a novice must overcome. The decision to choose non-specialists has other advantages as well. According to Pierre Samuel, a member of Bourbaki from 1947 until 1971, "It's easier to find a speaker." In addition, he adds, "We were thinking in terms of the educational aspect: it's useful for a young mathematician to talk about a subject he doesn't know especially well, it's a very fruitful exercise."

Of course, the quality of a seminar depends not only on the choice of speakers but on the choice of topics as well. In this respect, Bourbaki's choices have been sufficiently relevant to make the seminars renowned and well-attended. As Arnaud Beauville, who retired from Bourbaki in 1997, stated, "The Bourbaki seminars have become a meeting point for mathematicians from outside of Paris and outside of the country. Great effort is put into selecting the topics, and this is done so well that people trust Bourbaki's choices." The comments of Jean-Pierre Bourguignon, who was never a member of Bourbaki, supports this view. "When a subject is chosen for a Bourbaki seminar, it means that it is truly important and interesting," he asserted, adding that "it's one of the world's mathematical meccas. A young mathematician who is asked to give a lecture at a seminar must think twice about accepting. His reputation is at stake. It's exceptionally demanding intellectually, and the speaker is subject to great psychological pressure." And Bourguignon knows this well, as he was invited to speak on three occasions, in 1977, 1985, and 1991. Certainly the presence of the greatest mathematical minds of France, such as André Weil, Claude Chevalley, Henri Cartan, and Alexandre Grothendieck, in the front row of the lecture hall would be quite intimidating. "Some presentations didn't go well, and Dieudonné especially never let anything go unnoticed," Bourgignon explained. "And even though the front row is less prestigious today, it's still in the speaker's interest to thoroughly prepare his lecture."

Jean-Pierre Serre, a professor at the Collège de France, was one of Bourbaki's driving forces for many years.

Although the Bourbaki Seminar has attracted many participants and has become a mathematical institution—it even receives a small subsidy from the French government—it is also the target of some criticism. For example, the group has been particularly criticized for favoring certain areas of mathematics while completing ignoring others. This criticism, directed against Bourbaki's choices in general and not only against their choices of seminar topics, attacks Bourbaki's nearly exclusive interest in pure mathematics. Indeed, until the 1970s the lectures at the seminars mostly dealt with subjects in which the Bourbakis specialized, in particular, algebraic geometry (which flourished in the 1950s and 60s thanks to the work of French mathematicians), algebraic topology, and the theory of Lie groups. According to

Dieudonné's book *Panorama des mathématiques pures: le choix bourbachique* (published in English as *A Panorama of Pure Mathematics, as Seen by N. Bourbaki*), most of the subjects discussed in the Bourbaki Seminar dealt with "problems revolving around a general theory that is fruitful and alive" or with "problems generating new theories." Anything inspired by applied mathematics or other disciplines, such as physics or computer science, was more or less ignored. This simply reflects the individual tastes of the members of Bourbaki. The members, as far as we know, were all pure mathematicians who until recently were at most mildly interested in other sorts of mathematics. But the evolution of mathematics and its growing overlap with physics and technology have moved the group to change its attitude and even its tastes. Thus since the 1980s, the Bourbaki Seminar has been open to lectures on less pure subjects, of which Yuri Manin's lecture on quantum computing is one example.

Too Specialized, According to Some

Another, more disconcerting, criticism is that the lectures today are less educational than those of the 1950s. "The seminars are still very interesting, but they don't play the educational role they did in the past; they're becoming standard specialized seminars that are on the most part unaccessible by non-specialists," said Christian Houzel, a mathematician and math historian. Similarly, in an interview of three members of Bourbaki (called B1, B2, and B3) appearing in the journal *Gazette des mathématiciens* in January 1988, the mathematician Martin Andler writes that "The lectures dating from the 1950s seem very clear relative to those of today, which are far more complicated." According to Claude Chevalley's remarks appearing in Judith Friedman's 1977 dissertation, today people attend the seminars out of snobbery, and only very few actually understand the lectures. Chevalley confided that "as eminent a mathematician as Grothendieck ultimately decided to stop coming (long before he stopped doing math) because he felt he was wasting his time by attending lectures he didn't understand." It is certainly the case that these lectures never made for easy listening. While recalling memories about Dieudonné, Laurent Schwartz remarked in 1994 that "people came from all over France. I think that just a third of the audience understood, although everyone claimed to. It was very nice, but the level was quite high; Dieudonné was one of those who understood everything, while I was one of those who didn't."

Alexandre Grothendieck, who was a member Bourbaki and considered a genius, isolated himself from the world three decades ago.

Meanwhile, Pierre Cartier believes that the Seminar lectures provide quite a good overview of mathematical topics but misses the days—about twenty or twenty-five years ago—when "most of the lectures were given by members of Bourbaki; today, the seminars are not as closely connected to the group's activities, they're not as focused." A quick count illustrates this point. Among the first 49 lectures, from December 1948 to May 1951, more than 20 were given by members of Bourbaki. Between November 1972 and June 1975, 24 of the 53 lectures were given by members of Bourbaki. However, out of the 50 lectures between June 1995 and June 1998, a mere six or seven of the speakers were members of the group. The involvement of Bourbaki members in their own seminar is not what it used to be.

Overall, however, the Bourbaki Seminars are greatly appreciated. As Jean-Pierre Bourguignon said, "They taught me an enormous amount of math, and I regularly refer to the lecture notes." He doesn't believe that the Seminar has suffered any decline. The number of mathematicians attending is still large, and the seminars have benefited from opening up to subjects outside of pure mathematics. Those who criticize the Seminar's resemblance to the multitude of other seminars and minicolloquiums organized around the world may really be criticizing the loss of the Bourbakian spirit. Scientific quality beyond all doubt is not enough. Now that the seminars are the only window into the activities of Bourbaki, the addition of spirit and originality is needed more than ever.

Bourbaki's Parents: the Hadamard and Julia Seminars

In France following World War I, the one place vibrant mathematicians could be found was the Hadamard Seminar. While mathematical seminars had existed in Germany as a way of working and teaching since at least Carl Gustave Jacobi (1804-1851), the Hadamard Seminar was the first such seminar in France. It was formed in 1920 by Jacques Hadamard, a mathematician who was inquisitive about everything and who, according to André Weil, "retained an extraordinary liveliness of spirit and character until the end of his very long life." The Seminar existed until Hadamard's retirement in 1937. It was held one or two times a week at the Collège de France in Paris and was open to everyone. The Seminar's goal was to present a wide range of contemporary research in mathematics through discussing articles that Hadamard received from around the world. In his autobiography *Souvenirs d'apprentissage*, Weil recalls how his colleagues would meet at Hadamard's house in Paris's thirteenth

The Bourbakis respected Jacques Hadamard, who proved "the" theorem on the distribution of prime numbers in 1896.

arrondissement at the beginning of each year. There, Hadamard (who was open to suggestions for topics from the others) would distribute the articles to analyze and set the dates for the corresponding lectures. As for the seminars themselves, they were nothing like the numerous and unfortunately indigestible conferences called seminars today. "Hadamard acted as if the main goal of the Seminar were to teach him, Hadamard. It was to him that one addressed oneself, for him that one spoke. [...] When a lecture became unclear, he often asked for clarification or even gave one himself. At the end he would also make some comments, sometimes a few words, sometimes an extensive discussion. He never seemed aware of his superiority [...]" The audience also participated.

The Hadamard Seminar welcomed both novices (Weil was a doctoral student when he first participated in 1922) and established mathematicians (such as Émile Borel, Élie Cartan, Paul Lévy, and Gaston Julia), both French scientists and foreign ones (such as Sergei Bernstein, George Birkhoff, Vito Volterra, and Edmund Landau). This international forum of mathematics in France "was a meeting place for all mathematicians in Paris. Any mathematician visiting Paris from abroad was almost required to visit the Seminar," Liliane Beaulieu writes in her thesis. The Hadamard Seminar certainly contributed significantly to shaping the future founders of Bourbaki.

Indeed, these young mathematicians decided to form their own seminar in 1933. Weil, Chevalley, and their friends asked Gaston Julia, the youngest of their professors at l'École Normale, to sponsor the conferences they wanted to organize—Julia's help was necessary for obtaining a conference room, which had to be requested from the University of Paris. The Julia Seminar, which was simply called the *Séminaire de Mathématiques* until 1938, was held every other week on Monday afternoons in the Darboux Auditorium at the Henri Poincaré Institute. Unlike the Hadamard Seminar, each academic year was devoted to one theme (groups and algebras in 1933–34, Hilbert spaces in 1934–35, topology in 1935–36, Élie Cartan's work in 1936–37, algebraic functions in 1937–38, calculus of variations in 1938–39). These were not strictly current topics, but the Seminar's goal was to focus on a certain area and teach the main points of each topic. Every lecturer had to write down his lecture, and about forty copies of each lecture were distributed. The audience of the Julia Seminar was much smaller than that of the Hadamard Seminar. The nucleus of the group consisted of the Bourbakis (or, before 1935, the future Bourbakis) and their friends. Gaston Julia and Élie Cartan were regular participants, and some foreign mathematicians, such as John von Neumann and Carl Ludwig Siegel, occasionally attended. The Julia Seminar, where the Bourbakis first worked together in overhauling mathematics, didn't survive the war. But although the Julia Seminar came to an end in 1938, it was revived in 1948 in a different form: the Bourbaki Seminar.

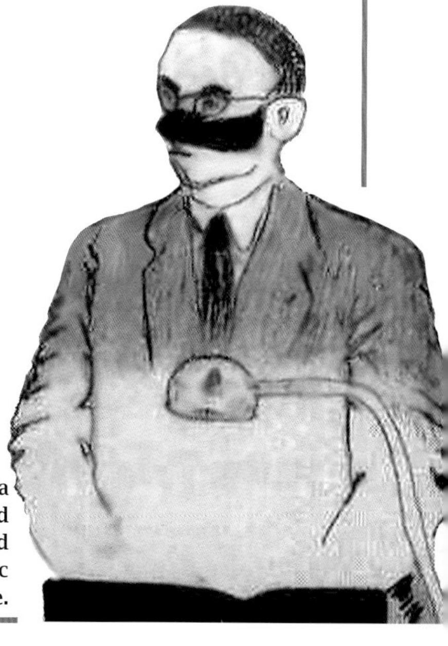

Gaston Julia (1893–1978), who had been wounded in World War I, wore a prosthetic leather nose.

8

Subtle and Austere Schoolboys

Like Janus, Bourbaki has two faces. The first, a public one, is serious and dry; the second, a private one, is humorous and joking. Between the two, Bourbaki's subtle spirit has done well.

It's hard to imagine that the author of *Éléments de mathématique* could be ready to laugh and joke. Refined, serious, dry, severe, impersonal: such is the style of Bourbaki's treatise. At first sight, the only fanciful element is the large, rounded Z placed in the margin wherever a reader could fall into a mathematical trap. A more attentive reader would discover strange mathematical terminology, which is familiar to mathematicians today but seemed new and fanciful in the fifties. While this suggests a certain humor, it still conceals the fact that behind Bourbaki was a group of real mathematicians who liked to have fun while working hard. To notice this essential aspect of Bourbaki's social life, it is necessary to look beyond the treatise, to look instead at the issues of *La Tribu*, the group's internal newsletter, copies of which were occasionally leaked to the outside world. Its here that the folklore of Bourbaki truly reveals its splendour.

The Weight of Words, the Crash of Barrels

Filters and ultrafilters; surjective, injective, and bijective functions; separated spaces and Polish spaces; barrels; inductive and projective limits; balls;...In the process of renovating mathematics, Bourbaki introduced a plethora of new words, either freshly invented or borrowed from everyday language, to mathematical terminology. Clarification of terminology inevitably went hand in hand with the clarification of concepts Bourbaki wished to achieve, and obviously this clarification of terminology sometimes demanded the creation of new words. Of

One-, two-, and three-dimensional balls.

course, it was necessary to do this elegantly and efficiently, by seeking simplicity rather than inventing heavy and pedantic jargon. In his 1968 lecture in Romania, Jean Dieudonné said that Bourbaki had abolished certain terms while inventing many others, "by using Greek when necessary, as everyone does, but also by using many words from everyday language. This has made many traditionalists bristle, as they had difficulty accepting the fact that instead of talking of *parallélotopes* and *hypersphéroïds*, we talked of *pavés* ['paving stones'] and *boules* ['balls'] [...]. Bourbaki's treatise is written in this style, rather than in jargon strewn with abbreviations, which are so common in English texts where they talk about the C.F.T.C, which is connected to an A.L.V., as long as it isn't a B.S.F. or a Z.D., and so forth. After ten pages of this you get completely lost. We believe that ink is not so expensive that it's completely impossible to write things out with well-chosen vocabulary."

More than thirty years later, the validity of Dieudonné's criticism is striking. Many of today's jargon-creators would do well to follow his advice. And indeed, Bourbaki's success in reforming terminology is great. If a prize for defending the French language had existed during the golden age of Bourbaki's treatise, the group certainly could have aspired to the award. For example, Christian Houzel remarked that it was Bourbaki who created the words *recouvrement* and *revêtement*, which name two distinct concepts falling under a single name in English ("covering"). "We made enormous efforts to come up with good terminology, and each new term had its noun, its adjective, and so forth," explained Michel Demazure, who was an active member of Bourbaki until around 1985. Bourbaki pushed for linguistic purity to the extent that they banned hybrids of Greek and Latin roots,

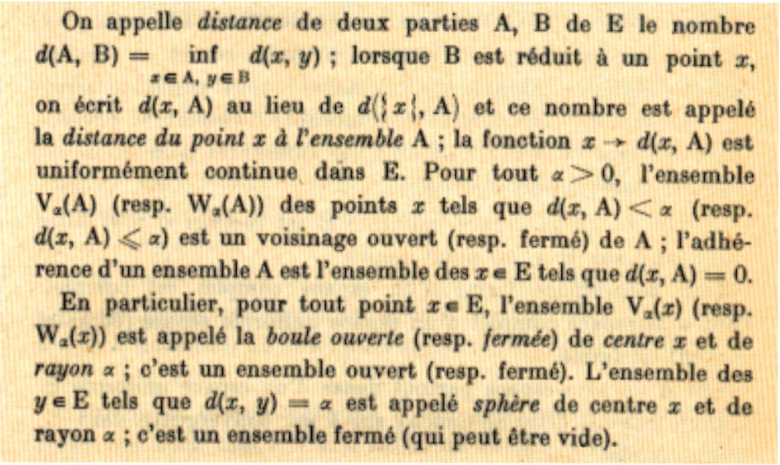
On appelle *distance* de deux parties A, B de E le nombre $d(A, B) = \inf_{x \in A, y \in B} d(x, y)$; lorsque B est réduit à un point x, on écrit $d(x, A)$ au lieu de $d(\{x\}, A)$ et ce nombre est appelé la *distance du point x à l'ensemble* A ; la fonction $x \to d(x, A)$ est uniformément continue dans E. Pour tout $\alpha > 0$, l'ensemble $V_\alpha(A)$ (resp. $W_\alpha(A)$) des points x tels que $d(x, A) < \alpha$ (resp. $d(x, A) \leqslant \alpha$) est un voisinage ouvert (resp. fermé) de A ; l'adhérence d'un ensemble A est l'ensemble des $x \in E$ tels que $d(x, A) = 0$.

En particulier, pour tout point $x \in E$, l'ensemble $V_\alpha(x)$ (resp. $W_\alpha(x)$) est appelé la *boule ouverte* (resp. *fermée*) de *centre* x et de *rayon* α ; c'est un ensemble ouvert (resp. fermé). L'ensemble des $y \in E$ tels que $d(x, y) = \alpha$ est appelé *sphère* de centre x et de rayon α ; c'est un ensemble fermé (qui peut être vide).

The extract from the book on general topology in which the world *boule* ("ball") is used for the first time in mathematics.
(For English translation, see appendix.)

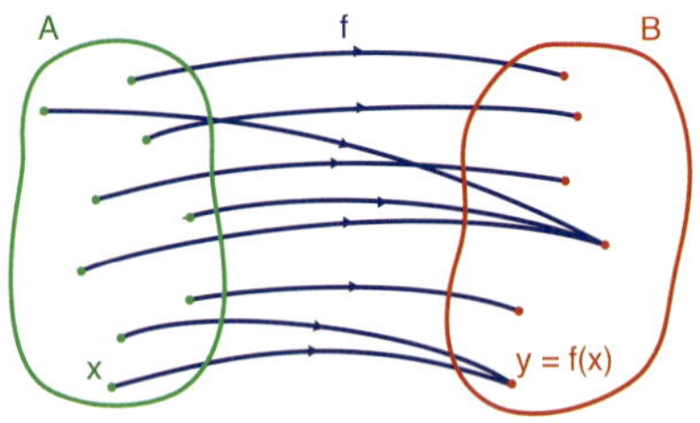

A function f defined from A to B is surjective if for every y in B there is at least one x in A that is mapped to y by f.

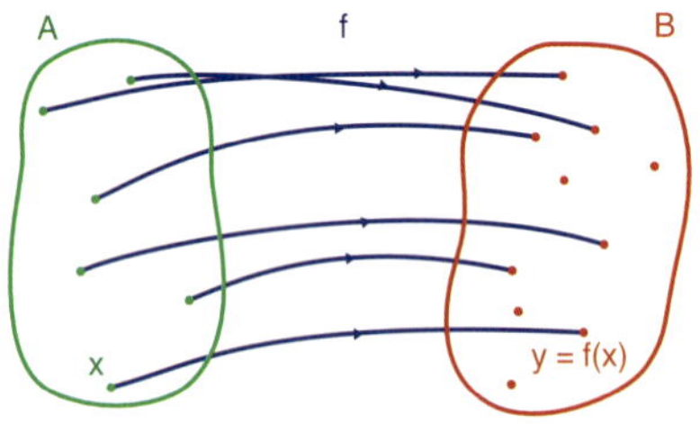

A function f defined from A to B is injective if for every y in B there is at most one x in A that is mapped to y by f.

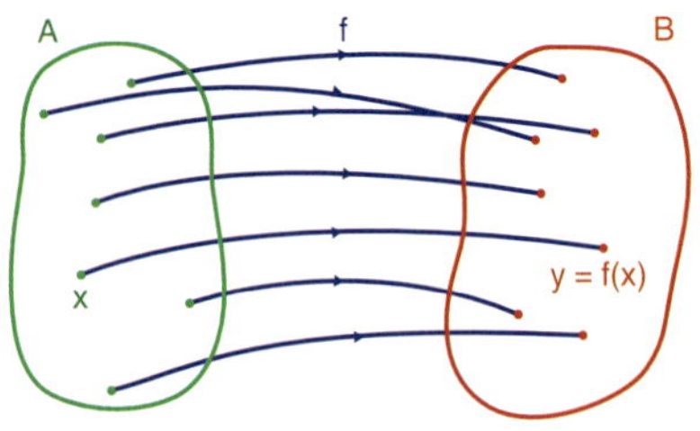

A function f defined from A to B is bijective if for every y in B there is exactly one x in A that is mapped to y by f.

forbidding words like "isojection" and "equimorphic." "André Weil heavily emphasized this," recalls Pierre Samuel. It is almost unanimously agreed that Bourbaki's efforts in terminology were worthwhile. Even though the group's lexical innovations met some opposition, many of the new terms ended up being adopted into French mathematical language. And after translation, the new words were introduced into other languages as well, including English and German.

Let's consider one of the examples mentioned above, the term *boule* (or "ball," its English translation). In three-dimensional space, a ball with center O and radius r is the set of points for which the distance to O is strictly less than r. (One can also talk about balls in n-dimensional space, with the definition being exactly the same.) By introducing this term, Bourbaki removed the amibiguity of the word "sphere", which previously could refer to either the container or its contents. After the introduction of the word "ball," the sphere with center O and radius r came to represent only the points whose distance to O equals r. In other words, a ball represents the volume, while a sphere is only the surface.

As a second example, consider the triplet surjective–injective–bijective, which was taught to schoolchildren during the era of New Math in the late 1960s. A surjective function (also called a surjection) from a set A to a set B is a function f defined on A such that any element y of B is the image of at least one element x of A. In other words, for any y in B there exists at least one x in A such that $y = f(x)$. A function is injective if each element y of B has at most one pre-image x in A; equivalently, f is an injection if the images $f(x_1)$ and $f(x_2)$ in B are different whenever x_1 and x_2 in A are different. Thus the application $f : \mathbf{R} \to \mathbf{R}$ defined by $f(x) = x^3$ is injective, since if x_1 and x_2 are different, then x_1^3 and x_2^3 must be different as well. On the other hand, the function $g : \mathbf{R} \to [-1, 1]$ defined by $g(x) = \sin x$ is not injective, since several different values of x (in fact, infinitely many) yield the same value of $g(x)$: for example, for any integer n the value of $\sin(n\pi)$ always equals 0. Finally, a function $f : A \to B$ is bijective if it is both injective and surjective, which means that for any y in B there is exactly one x in A such that $y = f(x)$. For example, the function $f : \mathbf{R} \to \mathbf{R}$ defined by $f(x) = x + 5$ is clearly a bijection. Meanwhile, the function $g : \mathbf{R} \to [0, +\infty)$ defined by $g(x) = x^2$ is not a bijection, since g is not injective (for example, $2^2 = (-2)^2$). The concepts of surjection, injection, and bijection are fundamental in characterizing functions from one set to another, so it's useful to have both a noun and an adjective corresponding to each concept. The prefixes sur-, in-, and bi- chosen by Bourbaki are common enough that it's easy to remember the definitions. Because of these qualities, these terms have been universally adopted in French,

At la Messuguière (near Grasse) in July 1975: Michel Demazure, François Bruhat, and Michel Raynaud.

and only somewhat less in English.

Not all of Bourbaki's neologisms achieved such posterity. As an example, Arnaud Beauville recalled the picturesque term *algèbres stellaires* ("stellar algebras"), which was suggested as an alternative to the term C^**-algèbres* (pronounced in French as the unweildy "C-star-algèbres," using the English "star" instead of the French "étoile"). Also, Pierre Samuel mentioned the adjective *étranger* ("foreign"), which failed to replace the phrase *premiers entre eux* ("relatively prime") used to describe integers whose only common divisor is 1 (for example, 10 and 21 are relatively prime).

While some of the new words, such as *surjectif*, *endomorphisme*, and *paracompact*, seem very learned, many others, such as *filtre* ("filter"), *pavé* ("paving stone"), *germe* ("sprout"), *encombrement* ("bulk"), and *concassage* ("a crushing"), all come from everyday language. They also suggest visual images of the concepts they define, which sharply constrasts with the characteristic dearth of diagrams and figures in *Éléments de mathématique*. It is through these sorts of words that Bourbaki's humor sometimes shines. A famous example is the term *tonneau* (translated and adopted in English as "barrel"), introduced in the group's book on topological vector spaces. Bourbaki was particularly happy to come up with this word, since a mathematical barrel is a "closed, convex, balanced, absorbing subset," with all of these adjectives, of course, having precise mathematical meanings. Meanwhile, on a more imaginative level, this term evokes barrels of spirits, echoing the prominence of the bottle of wine, armagnac, or champagne at Bourbaki conferences. As Pierre Cartier writes, "[Bourbaki] was so proud of this invention that reciting its definition became part of the

Espaces tonnelés.

1. Dans un espace localement convexe E, on appelle *tonneau* tout ensemble convexe équilibré, fermé et absorbant. Un tonneau absorbe toutes les parties convexes bornées et complètes (chap. III, § 3, *lemme* 1). Si E est quasi-complet, un tonneau absorbe toutes les parties bornées de E.

An extract from the book on topological vector spaces in which the word *tonneau* ("barrel") is used.
(For English translation, see appendix.)

ritual role-play punctuating the group's meetings [...] and forging its identity." Cartier also mentions another humorous aspect of Bourbaki's vocabulary: "Bourbaki's barely veiled contempt for certain subjects, which the group nevertheless had to discuss in its books, shows itself through unorthodox terminology. For example, in measure theory—a close relative of probability theory—the innocuous σ-algebras became fearsome 'clans, tribes, and phratries.'"

Was Bourbaki Intentionally Dyslexic?

Raymond Queneau points out in his book *Bords* that Bourbaki's treatise even contains an instance of word play: instead of writing the correct *ensemble filtrant à gauche et à droite* ("set filtering on the left and on the right"), Bourbaki writes *ensemble flirtant à gauche et à droite* ("set flirting on the left and on the right"). This famous misprint had escaped the eagle eye of Dieudonné, who almost always made the final revisions and checked the proofs, even long after he retired from the group. According to Laurent Schwartz in 1994, Pierre Samuel had in fact intentionally inserted this misprint with the editor's approval.

But Bourbaki's sense of humor is much more obvious in the issues of *La Tribu*, the group's internal newsletter. Subtitled *Bulletin oecuménique, apériodique, Bourbachique* ("An ecumenical, aperiodic, Bourbakian newsletter"), the newsletter's first issue dates from 1940. *La Tribu* was the successor of the *Journal de Bourbaki*, which was published between 1935 and 1937 and edited by Jean Delsarte, who was the group's secretary at the time. The main purpose of *La Tribu* is to report on the Bourbaki conferences. The first section includes a light and general summary of the conference, while the remaining part discusses technical decisions about the various drafts in progress. The first section is the most surreal, the most embelished with word games, Normalien slang, and humorous allusions to events (hilarious or otherwise) that took place during the conference. The issues of *La Tribu* provide a good taste of the legend Bourbaki created and of the atmo-

In addition to the regime of the High Commission, the terrible schism between mountaineers and couch potatoes threatens Bourbaki. Faced with an alpine valley, one took fright of the snow and made a dash for the Tropics; another rebelled against "these terrible mountains, these enormous masses lacking formality and structure"; a third, motorized, couldn't understand why the mountaineers insisted on being led to the bottom of each valley and abandoned them to meet their sad fate. Meanwhile, a delegation representing all ages and ranks set off to survey glaciers and neves, brave crevasses and mountain sickness, and plant Bourbaki's flag near the Caron shelter at 3,160 meters. The mountain air and high-spirited discussions keep Weil wide-awake and virulent. Professor Cartan, "seized by debauchery," surpassed himself in lewd wordplay.

An extract from *La Tribu* (Pelvoux-Le Poët conference, 1951).

sphere reigning over the group during its conferences. In addition, these documents help historians to trace the development of the various chapters of *Éléments de mathématique.*

Issues of *La Tribu* open like the script of a play, and Liliane Beaulieu remarks in her 1998 article that the narratives in *La Tribu*, especially those from the fifties and sixties, resemble the revues that the Normaliens produced each year for fellow students, former students, and special guests. All the issues of *La Tribu* follow the same basic structure. Issue 37, for example, opens with the title *Compte rendu du congrès de la lune* ("Report on the Conference of the Moon"), with the conference being baptized according to the editor's whim. Then all the conference's participants are listed ("Borel, Cartan, Chevalley, Godement, Grothendieck, Koszul, Schwartz, and the Blackboard") as well as the guinea pig (Malgrange). Issue 38 additionally includes a section titled "Miscellaneous" that listed all the necessary props: the blackboard, an eraser, three cars, forty-seven historical monuments, a gate. This is followed by a succinct table of contents, a narrative text mixing truth and fantasy, and even poetry. As issue 37 described,

On the Glacier Blanc, July 4, 1951: Laurent Schwartz, André Weil, Pierre Cartier, Pierre Samuel, Jean-Pierre Serre, and their guide. In the background is the Barre des Écrins, a peak of 13461 feet.

"The conference was held at the Mare aux Canards in Marlotte, near Fontainebleau, from October 23 through October 29, 1955.

"Modest and subdued sunlight, lustrous bronze leaves fluttering in the wind, a pond without fairies, modules without end, indigestible stones, and pierced barrels: everything contributes to the drowsiness of these blasé believers. 'Yet they are serious', says the hotel-keeper, 'I don't know what they are doing with all those stones, but they're working hard. Maybe they're preparing for a journey to the moon [...]."

This short extract illustrates several characteristics of the prose Bourbaki wrote for internal purposes. They often inserted mathematical terms (like "modules") into common language, alluded to questions and jokes from the conference, used vocabulary specific to the brotherhood (for example, *pierres*—meaning "stones"—referred to voluminous and problematic writing), likened the group to "believers" gathered around their great Master, pretended to belittle the importance of effort and work ("the drowsiness of the these blasé believers"), and putting emphasis on funny and festive events (with the "pierced barrels" clearly alluding to drinking).

Later in this same issue of *La Tribu*, we read that "in spite of Koszul's warnings, we managed to finish all the drafts that day. This earned us severe reprimands on the part of the Federation of Grape Pickers and Armagnac Distillers, who enforce a decree requiring us to consume 6 liters of alcohol per person per day [...]" We also read that "Chevalley did a wonderful act about automobile taming. A quadroon of believers set off blindly through the woods and across the fields, and, to their amazement, miraculously found themselves in the Marlotte town center. We were pleased to certify our adjudant's virtuosity in boat racing. Near Nemours, Schwartz noticed a placard announcing the distribution of currents." To fully understand Bourbaki's humor, it is necessary to know some mathematics. For example, the note about Schwartz is not funny unless you know that Schwartz created the theory of "distributions" and that "currents" are mathematical objects relating to this theory.

The text continues with a list of a few memorable quotes from the conference, such as "I'm not crazy in the way you think I am," "I remember which is which, but I always think of the other first," and "We need to make a cross-reference to the place where this isn't proven." Somewhat further, under the heading "Status of the Drafts," we read that this status is "stationary, with prevailing winds of hesitation, beautifully clear skies over Algebra, storms over the varieties, morning fog above Local Algebra, and royalties dropping noticeably." Next comes a more lengthy and serious section, with a statement of the tasks for the future, the schedule and approximate dates of the next conference, a few words on the upcoming Bourbaki seminar, and finally a detailed report on the current drafts and revisions to be made.

Throughout its issues, *La Tribu* is full of puns and intentional spoonerisms. For example, the phrase *le radon de la Mesure* ("Mesure's radon") appears, calling to mind the *radeau de la Méduse* ("Medusa's raft") while referring to the Radon measures used in the theory of integration. Another example is writing *Patrice de massage* (the "massage Patrice") instead of *Matrice de passage* ("Transition matrix"), which is an object in linear algebra. Bourbaki also filled *La Tribu* with Normalien slang (such as the verb *canuler*, derived from *canular*, meaning "a prank"), saucy statements (for example, "All the appendices are hung from Samuel's"), and pastiches of common expressions (such as "he who sows the functors reaps the structure"). The Bourbakis also invented words to use only among themselves, such as *croupion*, which is a subgroup of Bourbaki; *redactor demerdatur*, suggesting that the writer (the *redactor*) of a chapter was left to get himself out of any *merde*, or trouble; and *Diktat*, the name for a summons to a Bourbaki conference. Another amusing expression, invented by Pierre Samuel, is *l'âne qui trotte* ("the trotting donkey"), used to describe a mathematical development that is easy and "does all the work itself." Bourbaki even considered labelling easy exercises in its treatise with the symbol of a donkey! Another element of Bourbaki's unusual vocabulary was words derived from English, which were probably introduced thanks to the members' visits to America. Thus Bourbaki frenchified "delayed," "postponed," "disrupted," and "intricate" to create *délayé*, *postponé*, *disrupté*, and *intriqué*. Bourbaki also wrote poems, like the one André Weil wrote during the Chançay conference of September 1937, describing it as "a pretty faithful account of one of our discussions." A second poem was published in the Issue 8 of *La Tribu*. Written by Pierre Samuel, this poem is a pastiche of Mallarmé's poem *Le Cygne* ("The Swan").

Pierre Samuel and Serge Lang discussing Zariski's "Main Theorem" during the June 1956 Bourbaki conference in Sallières.

Bourbaki's Self-Irreverance

Bourbaki often caricatured and mocked its own members. For one thing, many Bourbakis were given a nickname for use within the group. Dieudonné was called by the anglified Dewdon, as a reminder of his visits to America and, perhaps, as a reference to the illustrious Isaac Newton. He was also called "the canonical adjudant," as he often urged his friends to work and his bursts of anger and shouting were memorable. Koko was the name for Jean-Louis Koszul, while Samuel Eilenberg had at least three nicknames: usually just Sammy, but also Zouly-Tapis and sometimes Heil Lemberg. Henri Cartan was a mosquito and a backseat driver (in reference to his cutting remarks), Jean Delsarte was a bishop, and so forth. Issue 39 of *La Tribu* (reporting the Conference of the Carpets, held in Sallières-les-Bains in June and July 1956) contains a not completely far-fetched caricature of some Bourbaki members based on a list of typical French personality traits published in the newspaper *France-Soir* (see opposite).

Bourbaki also made fun of relations among its generations—that is, among the "founding members," the "middle members," and the "lower members." For example, Issue 39 of *La Tribu* (reporting on the Conference of the Three Straight Angles, held from May 11 through May 17, 1956 in the Hôtel de la Brèche in Amboise) includes the report that "once again, Bourbaki ran into trouble with the authorities for breaking and entering private property. He was saved from the dungeon's damp straw only by the vehement intervention of one of the

Resourceful	**Schwartz**
Grumbling	**Godement**
Free-spirited	**Koszul**
Intelligent	**Ø**
Courageous	**Ø**
Hard-working	**Dewdon**
Greedy	**Sammy**
Individualistic	**Chevalley**
Patriotic	**Cartan**
Talkative	**Cartier**
Economical	**Dixmier**
Domestic	**Ø**
Skeptical	**Delsarte**
Frivolous	**Serre**
Sentimental	**Lang**
Logical	**Grothendieck**
Level-headed	**Samuel**

Sorbonne's most distinguished professors. We hope that he will promptly stop acting like a young hooligan. We will be forced to expel anyone who is not a Professor of the First Rank if he doesn't. Meanwhile, the regulations have been completed by articles requiring members to wear wing collars and ties, to use the title *Monsieur* when talking to undisputed leaders of the Sorbonne or the Collège [de France]. Also, members must address each other using the formal pronoun *vous*, and all guinea pigs must use the third person when speaking to their elders."

This sort of call for the respect of elders is absurd, since in reality no hierarchy or signs of deference existed within Bourbaki. In fact, this excerpt parodies and ridicules the sort of things that happened outside the group. Within the group, all members acted completely informally (although it appears that Weil was respected somewhat above the others). They made fun of each other without reserve, they played pranks, and sometimes they called each other bitter names during mathematical discussions. Yet the storms never lasted, and the group's cohesion was always strong. One anecdote Laurent Schwartz tells in his autobiography illustrates the violence of the group's discussions: "During a mathematical conference held in late June at Pelvoux-Le-Poët, a town in the Alps, we were the only guests in a small hotel run by a charming Mrs. Roland. [...] A violent dispute burst out over some question about mathematics. In a heated moment, André Weil slapped Henri Cartan on the head with a draft of a manuscript. The situation seemed hopeless. This frightened poor Mrs. Roland, who whispered to me, 'I suppose it's all over, that you'll be leaving?' I asked her why she thought so. 'You argue so much that you'll certainly going to split up.' I reassured her [...]. It goes without saying that peace was restored within ten minutes."

Agreeing with Weil's 1961 article titled *Organisation et désorganisation en mathématique* ("Organization and disorganization in mathematics), Armand Borel suggested that the anarchism of the group's discussions, and even their violence, was intentional. According to Borel, Weil believed that relatively new ideas have a better chance of being born in a confrontation than in an orderly discussion. "When [these ideas] emerged, the Bourbakis would say that *l'esprit a soufflé* ['the spirit shines forth'], and it's a fact that it shone much brighter after a 'spirited' (or perhaps I should say stormy) discussion than after a calm one," Borel explains.

One of the jokes the Bourbakis would play consisted of provoking Dieudonné to become beside himself with anger, causing his legendary—but always fleeting—resignation. Laurent Schwartz describes how "Dieudonné, who had a powerful voice, dominated the discussions

Pastiche par P. Samuel du poème «Le Cygne» de Mallarmé (1945)

Le Filtre

Ô puissant, ô formel, ô toi clair Bourbaki,
Vas-tu nous déchirer dans un accès de crise
Le Goursat filandreux, miroir de l'Analyse,
Défenseur attardé d'un passé qui a fui ?

La suite d'autrefois se croyait l'infini,
Inutile, et que sans la comprendre utilise
Le maladroit conscrit, lui que Valiron grise
De son cours ténébreux qui distille l'ennui.

Ignorant les secrets de la Topologie
À l'espace infligée, et toi qui l'études,
Il nage dans l'erreur où son langage est pris.

Il contemple étonné, comme enivré d'un philtre,
L'adhérence, un manteau qu'il n'a jamais compris,
Que vêt, sur un compact, immobile, le FILTRE.

(For English translation, see appendix.)

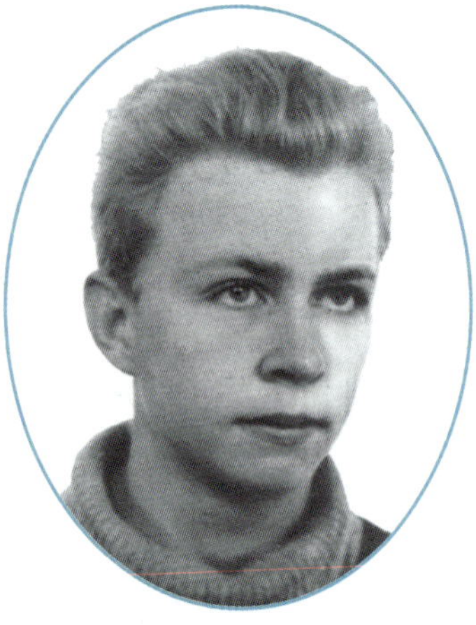

Adrien Douady

Adrien Douady reading a manuscript at a Bourbaki conference. Notice the string attached to his foot to prevent the manuscript from blowing away.

at the conferences. He regularly announced his thundering resignation. This didn't mean much, and in general, he would just continue on as before. [...] Some topics would make him resign automatically. For example, he wanted topological vector spaces to appear before integration, since he wanted to use them in explaining the theory of integration. As soon as anyone took it to his head to propose putting integration before topological vector spaces, Dieudonné would announce his famous resignation. Sonia Godement, the wife of the mathematician Roger Godement, was sceptical and wanted to see such a scene for herself. We invited her to come to the next meeting at exactly ten o'clock, the hour when we decided to incite the usual resignation. At three minutes before ten, someone calmly suggested, 'It's absolutely necessary to put integration before topological vector spaces.' Dieudonné resigned violently on the spot, at the exact moment when Sonia Godement arrived."

Another prank among the members was called *le cocotier* ("the coconut tree"). According to Liliane Beaulieu, this was inspired by a Polynesian custom where an old man climbs a palm tree and holds on tightly while someone shakes the trunk. If he manages to hold on, he remains accepted in his social group. Bourbaki translated this custom as the following: some members would set a mathematical trap for the others. If someone fell for it, they would yell out *cocotier*!

Bourbaki Marries Off His Daughter and Dies

But many of Bourbaki's pranks were intended for people outside the group. Some of the ones relating to the name Bourbaki are described in Chapter 2. Here is another, which is a real gem of Bourbaki history: the announcement of the marriage of Betti Bourbaki to Hector Pétard. The character Pétard was born in 1938, when a group of mathematicians from Princeton led by Ralph Boas and Frank Smithies were

Sonnet composé par André Weil au congrès de Chançay (septembre 1937)

Soit une multiplicité vectorielle.
Un corps opère seul, abstrait, commutatif.
Le dual reste loin, solitaire et plaintif,
Cherchant l'isomorphie et la trouvant rebelle.

Soudain, bilinéaire, a jailli l'étincelle
D'où naît l'opérateur deux fois distributif.
Dans les rets du produit tous les vecteurs captifs
Vont célébrer sans fin la structure plus belle.

Mais la base a troublé cet hymne aérien :
Les vecteurs éperdus ont des coordonnées.
Cartan ne sait que faire et n'y comprend plus rien.

Et c'est la fin. Opérateurs, vecteurs, foutus.
Une matrice immonde expire. Le corps nu
Rentre en lui-même, au sein des lois qu'il s'est données.

(For English translation, see appendix.)

inspired by Bourbaki's existence (which they learned of from Chevalley and Weil) to play their own prank. Adopting the pseudonyms Pétard and Ersatz Stanislas Pondiczery, this group with no real mathematical objective published an article about the mathematics involved in lion hunting in the *American Mathematical Monthly*. This famous lion hunt inspired a whole series of articles published between 1965 and 1985. In the spring of 1939, André Weil met Ralph Boas and Frank Smithies during a trip to Cambridge, England. With his wife Éveline and Claude Chabauty, a new member of Bourbaki, Weil decided to write and print an announcement of the marriage of Betti Bourbaki to Hector Pétard. This document, which was sent to all Bourbakis, mathematicians at Princeton, and others, parodies the announcements circulated in posh circles, where the name of each person is followed by a whole litany of honorific titles. It abounds with mathematical terms outside their scientific context ("uniform," "filter," "induced structure," "class-field," "covering," "well-ordered," "modulo," "left ideals," etc.). Betti numbers, named after Enrico Betti (1823–1892) are used in algebraic topology.

A similar prank was played in 1968, but this time Bourbaki was its target. The anonymous author (who turned out to be the author-mathematician Jacques Roubaud, a member of the Oulipo literary movement), wrote an announcement of Bourbaki's death, imitating the style of the marriage announcement. This document, which certainly

The announcement of Betti Bourbaki's marriage to Hector Pétard.

Monsieur NICOLAS BOURBAKI, Canonical Member of the Royal Academy of Poldevia, Grand Master of the Order of the Compacts, Curator of the Uniforms, and Lord Protector of the Filters, and Madame (*née* BIUNIVOQUE) have the honor of announcing the marriage of their daughter BETTI to Monsieur HECTOR PÉTARD, Administrator-Delegate of the Society of Induced Structures, Full Member of the Institute of Class-field Archaeologists, Secretary of the Work of the *Sou du Lion*.

Monsieur ERSATZ STANISLAS PONDICZERY, retired Covering Complex of the First Class, President of the Rehabilitation Home for the Weakly Converging, Knight of the Four U's, Grand Operator on the Hyperbolic Group, Knight of the Total Order of the Golden Mean, L.U.B., C.C., H.L.C., and Madame (*née* COMPACTENSOI), have the honor of announcing the marriage of their pupil HECTOR PÉTARD to Mademoiselle BETTI BOURBAKI, former student of the Wellordered of Besse.

They will receive the trivial isomorphism from P. Adic, of the Order of the Diophantines, in the Fundamental Cohomology of the Universal Variety, on the third of Cartember, year VI, at the usual hour.

The organ will be played by Monsieur Modulo, Assistant Simplex of the Grassmannian (with lemmas sung by the Scholia Cartanorum). The collection will be donated in full to the retirement home for Poor Abstracts. Convergence will be guaranteed.

Following the congruence, Monsieur and Madame Bourbaki will receive guests in their fundamental domains. Music for the reception will be provided by the fanfare of the Seventh Quotient Field. Canonical attire.
(Left ideals in the bouttoniere) Q.E.D.

An anti-Bourbaki parody: the announcement of Bourbaki's death.

The Cantor, Hilbert, and Noether families;
The Cartan, Chevalley, Dieudonné, and Weil families;
The Bruhat, Dixmier, Godement, Samuel, and Schwartz families;
The Cartier, Grothendieck, Malgrange, and Serre families;
The Demazure, Douady, Giraud, and Verdier families;
The Right-Filtering and Strict-Epimorphism families;
Mesdemoiselles Adèle and Idèle;
regret to announce the death of Monsieur

NICOLAS BOURBAKI

Respectively their father, brother, son, grandson, great-grandson, and grand-cousin.

He died piously in his home on November 11, 1968 (on the anniversary of the great victory) in his home in Nancago.The burial will take place in the cemetery for Random Functions (metro stations Markov and Gödel) on Saturday, November 23, 1968 at 3 o'clock in the afternoon.

A reception will be held at the bar The Direct Products, at the crossroads of the Projective Resolutions (formerly Koszul Square).Following the wish of the departed, His Eminence the Cardinal Aleph I will hold a mass in Our Lady of Universal Problems in the presence of representatives from all the equivalence classes and from all the (algebraically closed) fields. The students from l'École Normale Supérieure and the Chern classes will observe a minute of silence.

No flowers or wreath products.

"For God is the Alexandrov compactification of the universe." Groth.IV.22

reflects its author's wishes or criticisms, is another gem of Bourbaki history.

It's clear that Bourbaki really liked (likes?) to laugh, to laugh at its own expense, to laugh by playing harmless schoolboy jokes and by poetically exaggerating insignificant events, to laugh by personifying "Master" Bourbaki surrounded by his "believers," and so on. The Bourbakian humor takes diverse, often irreverent and dry, forms. Its importance in the life of the group is certainly not a coincidence, and Bourbaki's personality existed for a reason. "No one would have dreamed to introduce an austere, rigid mathematician with no sense of humor," writes Laurent Schwartz. Furthermore, laughter definitely strengthened social bonds. It helped the group to forge an identity, to obliterate internal hierarchy, to pacify confrontations between members, to release intellectual tension created by the mathematical discussions. If Bourbaki didn't know how to laugh, perhaps the group wouldn't have been able to survive and wouldn't have produced a work of such importance. And without the humor and the myths surrounding Bourbaki, the group's charm would certainly have been less effective.

9

"For the Honor of the Human Spirit"?

Bourbaki didn't only receive praise. Indeed, it met a certain amount of bitter criticism for its style, its choices, its power, and its lack of interest in anything outside of its vision of pure mathematics. Some of this criticism came from the group's own members.

In a letter to the French mathematician Adrien-Marie Legendre in 1830, the German mathematician Carl Gustav Jacob Jacobi wrote the following memorable words: "[...] Fourier believed that the main goals of mathematics were to serve the common good and to explain natural phenomena. However, a philosopher like him should have known that science is solely for the honor of the human spirit. Therefore a question about numbers is as important as a question about the world." Quoted time and again, Jacobi's eloquent phrase has become an emblem for the defense of pure mathematics. Jean Dieudonné named the book he wrote for the general public *Pour l'honneur de l'esprit humain* after this phrase (although it is published in English as *Mathematics - The Music of Reason*, after a quotation by the nineteenth-century mathematician James Joseph Sylvester) and repeated Jacobi's words in his epigraph. It is not a coincidence that this eminent former member of Bourbaki chose to emphasize Jacobi's point of view. Bourbaki was—and still is—a group of pure mathematicians who are interested mainly in pure mathematics.

Carl Gustav Jacobi, a "pure" mathematician.

The fact that Bourbaki is anchored in pure mathematics—and even a specific type of pure mathematics—did not only create admirers. In fact, many mathematicians detested Bourbaki. The group's style, its choices and omissions, its global vision of mathematics, and its influence all gave rise to harsh criticism. Today this group no longer holds a prominent place in the mathematical world, and the heat of criticism has fallen. But some criticisms remain, and discussing them will help show Bourbaki's place in the world of mathematics.

First, it is not really just to criticize Bourbaki for confining itself to

Fernand Léger's *Still life with beer mug* (1921). In the early the twentieth century, abstraction grew in art as well as in mathematics. In both disciplines, this increased the effort required for understanding a work's features.

pure mathematics. After all, one can't do everything—to each his own tastes and work! Also, the group didn't claim to cover all of mathematics, as it undertook to write a treatise on fundamental subjects that would be useful for all mathematicians. Bourbaki didn't wish to

Adrien-Marie Legendre, initially an "applied" mathematician.

explain theories specific to this or that mathematical domain, so it would also be unjust to criticize Bourbaki for failing to do so. The aspects of Bourbaki's work that do deserve criticism lie elsewhere. The Bourbakis did have to make a choice of approach for each topic they discussed, since there are almost always several different ways of approaching and developing a mathematical subject. However, their choices for *Éléments de mathématique* were sometimes unfortunate. In addition, Bourbaki lacked interest, and sometimes even showed contempt, for any areas not directly connected to pure mathematics. This would not have provoked any anger if the group had not gained considerable intellectual, and even institutional, power. But thanks in part to the group's Seminar (see chapter 7), Bourbaki's influence was so strong that the group's views directed French mathematics a great deal, and some scientific areas suffered because of this. In retrospect, Bourbaki showed a certain narrowness of perspective relative to the evolution of world mathematics, even while the group unquestionably contributed to the rebirth of French mathematics and helped raise it to hold third or fourth place in the world today.

Bourbaki's Choice: Neither Logic nor Applied Math

What was Bourbaki criticized for exactly? The subjects that Bourbaki neglected, and even scorned, included all mathematics having to do with applications. Among these neglected subjects were numerical analysis (which is about numerical calculations and solutions to equations or to other sorts of problems), probability theory, theoretical computer science, game theory, and optimization theory. It is true that most of the topics in applied mathematics lend themselves poorly to the axiomatic method and structural vision that Bourbaki advocated. But Bourbaki's lack of interest in these areas held back their development in France, while applied mathematics in the United States and the Soviet Union benefitted from rapid development starting in the 1940s with the advent of World War II. Even Jean Dieudonné, a zealous member of Bourbaki, admitted in his old age the wrong done to applied mathematics. As he told Marian Schmidt in 1990, "It is possible to say that there was no serious applied mathematics in France for forty years after Poincaré. There was even a snobbery for pure math. When one noticed a talented student, one would tell him 'You should do pure math.' On the other hand, one would advise a mediocre student to do applied math while thinking, 'It's all that he can do!'" In skillful praise of pure mathematics, Dieudonné added that "the truth is actually the reverse. You can't do good work in applied math until you can do good work in pure math."

It is worth discussing the case of probability theory specifically, since Bourbaki's lack of interest in this subject led Bourbaki to choose an approach for one of the books of *Éléments des mathématique* that many today consider unsuitable. Andrei Kolmogorov's work around 1930 made the study of probability into a quite abstract mathematical theory. In particular, this Soviet mathematician developed a set of axioms that form a rigorous foundation for probability theory. This theory is closely linked to measure theory and integration theory, which were founded by Émile Borel and Henri Lebesgue in the early twentieth century (see the box on page 127). Yet Bourbaki decided against using Lebesgue's methods in discussing integration theory. Under the influence of André Weil, Bourbaki instead chose to use the concept of Radon measures on locally compact spaces, which have no use whatsoever in probability theory. As the mathematician and mathematical historian Christian Houzel explained, "For many topics—the study of Brownian motion for example—probability theory relies on a measure defined on sets that are not locally compact. In the end, Bourbaki decided to add a chapter about spaces that are not locally compact to the group's book on integration, but this was akin to admitting that the previous chapters were nearly worthless."

Maybe Bourbaki would have made a different decision about what approach to use if the group had been interested in probability theory, but this in fact was not the case. "Bourbaki stepped away from probability, rejected it, considered it to be unrigorous. The group's considerable influence directed young mathematicians away from probability as well. The Bourbaki group, including myself, is greatly responsible for holding back thc development of probability theory in France," Laurent Schwartz writes in his autobiography. Perhaps Bourbaki's contempt for probability is illustrated by what Schwartz writes about a lecture that the eminent American mathematician Joseph Doob gave in Paris some years ago. "The Bourbakis in the audience would interrupt constantly, loudly, and impolitely, saying that the space was not locally compact and that it 'didn't mean anything.' Their attitude was deeply troubling and their lack of courtesy outraging."

Andrei Kolmogorov (1903–1987), the father of modern probability theory.

Bourbaki's Lack of Interest in the Foundations

Neither was Bourbaki's attitude towards mathematical logic exemplary. At the time, it certainly would have been impossible to criticize the area for being too close to applied mathematics. But for Bourbaki, logic was something outside of mathematics. Jean Dieudonné, who liked to make cutting remarks, writes in the article *Mathématiques vides et mathématiques significatives* ("Empty mathematics and meaningful mathematics") that ninety-five percent of mathematicians "don't care a fig" for mathematical logic, that "when someone comes and talks to us about first and second order logic, recursive functions, and models, which are very nice and very beautiful theories that have yielded remarkable results, we, the mathematicians, don't see any reason people shouldn't study these topics, but on the other hand they don't attract us a bit." This nearsightedness is surprising in a mathematician who had such a broad education in his subject. Now the concepts he sarcastically mentioned play an important role in mathematics, particularly in connection with theoretical computer science. Of course, now these concepts are starting to have applications...

In his 1948 article *L'avenir des mathématiques* ("The future of mathematics"), André Weil expresses his hardly less caricatural point of view: "[...] logic is the mathematician's hygiene, but it does not provide his meals. The big problems constitute the daily bread on which he lives." In other words, Weil believes that none of the problems that logicians study are "big" problems in mathematics. Even the question of whether mathematics is internally coherent leaves Weil as unfeeling as marble. "It would be possible [...] that one day experience would lead us to discover the seed of contradiction in the methods of reasoning we currently use. Then we would need to revise mathematics. It is certain that this won't affect the heart of the subject." Here Weil is probably alluding to Kurt Gödel's proof from around 1930, which showed that it is impossible to prove that a system of axioms at the base of mathematics will never lead to a contradiction, if the proof uses only the axioms in question. Surprisingly, the spectacular results of Gödel and other logicians after him do not bother most mathematicians. They are satisfied with taking a pragmatic stand and not worrying too much about logical subtleties that could affect the foundations of their discipline. This view is understandable, as experience shows that questions about the foundations of mathematics rarely have repercussions in a mathematician's everyday work. That Bourbaki would adopt this point of view is less acceptable, since the group took pride in setting

Sierpinski's *Leçons sur les nombres transfinis.*

COLLECTION DE MONOGRAPHIES SUR LA THÉORIE DES FONCTIONS
PUBLIÉE SOUS LA DIRECTION DE M. ÉMILE BOREL

LEÇONS

SUR LES

NOMBRES TRANSFINIS

PAR

WACLAW SIERPIŃSKI

PROFESSEUR A L'UNIVERSITÉ DE VARSOVIE
MEMBRE DE L'ACADÉMIE POLONAISE DES SCIENCES ET DES LETTRES

PARIS

GAUTHIER-VILLARS ET Cie, ÉDITEURS

LIBRAIRES DU BUREAU DES LONGITUDES, DE L'ÉCOLE POLYTECHNIQUE

Quai des Grands-Augustins, 55.

1928

In la Messuguière in July 1975: Jean-Louis Verdier, Bernard Tessier, André Gramain, and Hyman Bass.

mathematics on the solid base of set theory, the axiomatization of which relies on formal logic.

Furthermore, the Bourbakis' lack of interest in mathematical logic rubbed off on their book on set theory, which starts with a chapter on formal logic. The Bourbakis themselves admit that this book was written "with pain and without pleasure, but we had to do it." Logicians disparaged the result. For example, in his article "The Ignorance of Bourbaki" published in *The Mathematical Intelligencer* in 1992, the British mathematician A.R.D. Mathias describes how Bourbaki's book on set theory shocked him profoundly: "It appeared to be the work of someone who had read *Grundzüge der Mathematik* by Hilbert and Ackermann, and *Leçons sur les nombres transfinis* by Sierpinski, which were both published in 1928, but nothing since." With reason, Mathias emphasizes that when Bourbaki and its members wrote about logic, they usually ignored Gödel and his work, and logicians believe this is unacceptable. A logician from Paris criticized Bourbaki at least as severely as Mathias did by saying that "these chapters should be thrown away [...] The Bourbakis wrote them even though they weren't interested in

A.R.D. Mathias criticized "the ignorance of Bourbaki."

Was Bourbaki too Powerful?

Did Bourbaki monopolize power in the French mathematical community? Bourbaki certainly held considerable intellectual power from the fifties through the seventies. It is also true that many members of Bourbaki held important positions in the mathematical community; they were university presidents, mathematics department chairs at l'École Normale Supérieure, presidents of the French Mathematical Society, presidents of the Centre International de Rencontres Mathématiques (in Marseilles), and members of the French Académie des Sciences. It's less clear, however, that Bourbaki's power exerted itself through these institutional positions. "Everything ascribed to Bourbaki's politics is false. Outside of mathematics, Bourbaki is a much too heterogenous group to be able to define a common position. During my twenty years in Bourbaki, I almost never heard anyone talk about university politics," asserts Michel Demazure. Jean-Pierre Kahane, who was never a member of Bourbaki, also believes that Bourbaki did not try to exert political power, saying that "Bourbaki had intellectual power, not institutional power. Its members didn't hold that many positions of power, and no one criticized people like Delsarte or Dieudonné for their use of power. In fact, they were known for their strict ethics." And Jean-Pierre Bourguignon judges that during the sixties Bourbaki was "a pressure group in the good sense of the word," as its members acted in the interest of the whole mathematical community.

However, these favorable opinions clash with those held by Claude Chevalley, one of the founding members of Bourbaki. In an interview with Denis Guedj published in 1981, Chevalley criticizes the role played by the group in the question of university careers. "From the first conferences until the War," he says, "it was tacitly understood that one should not discuss questions of university careers. Unfortunately, such discussions became pervasive after the War. This was probably because young mathematicians were entering the group, and naturally we wanted to make sure they would find jobs. This was the beginning of a fatal spiral.

Claude Chevalley and François Bruhat at the Bourbaki conference in Amboise, October 1956.

Little by little, we began talking about everyone's careers: it was complete decadence." He also criticizes the change in the group's attitude towards honorific titles. "We had agreed that no member would join the Académie [des Sciences]—I recall the conversations on the subject perfectly. But now they're almost all members."

Everyone agrees that Bourbaki's intellectual influence is strong, or even stifling. Jean-Pierre Bourguignon explained that he felt pressure to focus on algebraic geometry but decided to turn to differential geometry under the influence of Marcel Berger. It was only later, during a visit to the United States, that he realized that the French research in this area was not as poor as he had imagined. Marcel Berger and his students created a French mathematical school that existed alongside the Boubakian school. In addition, Bourguignon explained, an anonymous group similar to Bourbaki was formed around 1974. Named Arthur L. Besse, this group organizes seminars and has published four books on specialized subjects. Incidentally, the pseudonym Arthur L. Besse comes from the fact that the group called its meetings "round tables"—hence the first name Arthur, with L for Lancelot—and because the group held it first conference in Besse-en-Chandesse, the same place as Bourbaki's founding conference! In addition to Arthur L. Besse, several other centers of activity existed separately from Bourbaki, including Gustave Choquet's school of analysis, Jacques-Louis Lions' school of applied mathematics (formed in the early sixties), and André Lichnerowicz's school, which worked differential geometry in problems in theoretical physics. Jean Leray, a great mathematician who had almost no students, also remained separate from Bourbaki. Thus the Bourbaki group was not the only mathematical group in France, although until the seventies it was certainly at the top of the ladder. As Armand Borel, a former Bourbaki, wrote in 1998, "The mathematical atmosphere was not favorable to mathematicians who had a different temperament, a different approach. It's truly regrettable, but we can't hold it against the members of Bourbaki, who didn't force anyone to do research in their way."

the subject and weren't familiar with the work of contemporary logicians. They presented their own system of logic, which turns out to be unusable. The problem is that this book was treated as the main reference for mathematical logic for a long time in France. This caused a lot of damage to the field. The idea that logic is not an interesting subject spread, and traces of this view still remain in France's mathematical community today."

Bourbaki was also strikingly uninterested in physics. Over the course of the previous centuries, physics and mathematics had influenced each other greatly. Mathematics is very useful to physicists, and conversely, problems in physics have inspired many developments and discoveries in mathematics. For example, Fourier introduced his famous trigonometric series while studying the equations of heat propogation. Even the great turn-of-the-century mathematicians David Hilbert and Henri Poincaré were deeply interested in physics and contributed to

David Ruelle, a professor at the Institut des Hautes Études Scientifiques, was one of the pioneers of chaos theory.

clarifying certain topics in this discipline. This was also the case for Bartel van der Waerden and Emmy Noether, who were among the German algebraists so respected by the Bourbaki founders. But this was not the case for the Bourbakis, at least until recently. The example of André Weil is typical. Despite the fact that Göttingen, where Weil lived in 1926, was one of the centers of the quantum physics revolution during the 1920s, Weil didn't notice this development. "As I learned much later," he writes in *Souvenirs d'apprentissage*, "the world of physicists was bubbling with excitement in Göttingen as they were giving birth to quantum mechanics. It's quite remarkable that I didn't notice what was going on around me at all."

In Bourbaki's defense, it must be said that the decades between 1930 and 1960 were not very rich in mathematical physics, the area of physics most inclined to interact with mathematics. According to David Ruelle, a physicist-mathematician working at the Institut des Hautes Études Scientifiques near Paris, "The arrival of quantum mechanics is one of the reasons." Scientists put a massive amount of effort into this new field since there were so many new things to discover, but—at least at that time—it only required moderately elaborate mathematics. Moreover, many physicists showed contempt for the

Bourbaki style of mathematics, which was much more formal than what they were used to and what they thought they needed. These considerations could be what caused some scientists, such as Pierre-Gilles de Gennes (who received the 1991 Nobel Prize in Physics) and Claude Allègre (a geochemist and former French Minister of Education), to criticize mathematics. Meanwhile, mathematicians already had plenty of material to create their own problems, and the internal dynamics of mathematics moved most of the flourishing areas (especially algebraic geometry and algebraic topology) away from the cares of physicists. As Weil said in an interview with the *Gazette des mathématiciens* in October 1991, "Overall, it was a period where most of the big developments in mathematics did not originate in physics."

René Thom, the founder of catastrophe theory.

In short, Bourbaki was mostly impermeable to all subjects outside the heart of pure mathematics. But two remarks should be made to qualify this statement. First, it's important to differentiate Bourbaki from its members. For example, Claude Chevalley, who was the most philosophical of the group, became interested in logic thanks in part to the influence of his friend Jacques Herbrand. It was Chevalley who pressed the group to include formal logic in its treatise. He also wrote a long introduction to Bourbaki's book on set theory. Chevalley was very fond of this text, according to his daughter Catherine, but the group decided against its publication. Similarly, some members, including Laurent Schwartz, Jean-Louis Verdier, and Pierre Cartier, were very open to physics and applied mathematics. Schwartz, whose work affected physics and applied mathematics, gave a course on the mathematical methods of physics for several years. Meanwhile, Cartier, who was interested in the mathematical problems of applied physics, was the first to include lectures on mathematical physics in Bourbaki's seminars.

The second remark is that while Bourbaki sealed itself from neighboring areas of science, the converse was not true. Bourbaki's style and methods influenced researchers in applied mathematics, mathematical physics, and other fields. One example of this influence is the case of Jacques-Louis Lions (1928–2001), who founded the modern school of applied mathematics in France. This former Normalien, who entered l'École Normale in 1947 and studied under Laurent Schwartz, was never a member of Bourbaki. Nevertheless, the group's views show in the somewhat formal nature of his work. Another scientist influenced by Bourbaki is the economist Gérard Debreu, who won a Nobel Prize in 1983 for introducing methods of analysis into economics and for giving a rigorous reformulation of the theory of general equilibrium. Debreu is also a former Normalien, who entered in 1941 and studied mathematics. He turned towards economics after graduation and moved to the United States. Influenced by Bourbaki (he was taught by

Benoît Mandelbrot, the father of fractals.

Alexandre Grothendieck (presenting), with Jean Dieudonné to his left and Claude Chevalley to his right.

Henri Cartan), Debreu introduced the axiomatic method into economics. He thought that economics is not an experimental science like physics and thus that the internal logic of theoretical models in economics must be absolutely coherent. Hence, Lions believed, the axiomatic method is necessary in the theory of economics.

Measure, Integration, and Probability

The integral is a typical example of a mathematical concept that can be approached from several different points of view. To explain this, let's limit ourselves to the case of a function f defined on an interval $[a,b]$ and taking positive real values. The integral of f over this interval is denoted by $\int_a^b f(x)dx$, whose value corresponds to the area between the x-axis and the curve representing f—one could say that the integral of f "measures" this area. The goal of integration theory is to give a precise meaning to this for as wide a class of functions as possible. The oldest approach is that of Cauchy and Riemann, which basically comes down to defining the integral as the following. The interval $[a,b]$ is subdivided into N intervals of length $\Delta x_1, \Delta x_2, \ldots, \Delta x_N$, and then the sum

$$S_N = f(x_1)\Delta x_1 + f(x_2)\Delta x_2 + \cdots + f(x_N)\Delta x_N,$$

where each x_i is a point in the interval Δx_i, is calculated as N approaches infinity. In other words, the area in question is calculated by approximating the region with N rectangles and calculating the sum of the area of these rectangles as N approaches infinity (see box 7 on page 64).

The modern approach, which was developed by Lebesgue, relies on the help of measure theory. In simplified and unrigorous terms, a measure defined on the set $\mathbf{R}$ of real numbers is a function m that associates to every subset A of $\mathbf{R}$ a nonnegative number $m(A)$ such that $m(A_1 \bigcup A_2) = m(A_1) + m(A_2)$ whenever the subsets A_1 and A_2 are disjoint (that is, whenever they have no common elements). These properties of nonnegativity and additivity make measures correspond to the intuitive concepts of length, area, and volume. It can be shown that there is a unique measure m on $\mathbf{R}$ such that $m([a,b]) = b - a$ for all a and b (in other words, so that the measure of any interval is its length). This measure is called the Lebesgue measure.

Now, how can one use the Lebesgue measure to define the integral of a function f defined on an interval $[a,b]$ and taking real values? Instead of subdividing the interval $[a,b]$ of the x-coordinates as in the case for Riemann integrals, we subdivide an interval $[y_{min}, y_{max}]$ of the y-axis that contains all the values taken by the function f. This yields N consecutive intervals

$$I_1 = [y_{min}, y_1), I_2 = [y_1, y_2), I_3 = [y_2, y_3), \ldots,$$
$$I_{N-1} = [y_{N-2}, y_{N-1}), I_N = [y_{N-1}, y_{max}].$$

For each of these intervals I_p, let A_p be the set of x such that $f(x) \in I_p$. (For example, A_2 is the set of x such that $f(x)$ lies between y_1 and y_2.) In general, A_p is not a simple interval but rather a union of disjoint intervals, or even something more complicated. Now, consider the product $z_p m(A_p)$ for each p, where m is the Lebesgue measure and z_p is a point in I_p, and construct the sum

$$S_N = z_1 m(A_1) + z_2 m(A_2) + \cdots + z_N m(A_N),$$

which approximates the area below the curve representing f and above the x-axis. When the interval $[y_{min}, y_{max}]$ is subdivided into finer and finer intervals, which corresponds to letting N approach infinity, the above sum generally approaches some number. By definition, this sum is the Lebesgue integral of the function f over the interval $[a,b]$.

It would be necessary to delve into the definitions and theorems in detail to show why the Lebesgue integral has numerous advantages over the Riemann integral. Most importantly, however, the Lebesgue integral is more general, since every function that is Riemann integrable is also Lebesgue integrable (with the values of the two integrals being equal), while there are functions that are Lebesgue integrable but not Riemann integrable. For example, the function g defined by $g(x) = 1$ whenever x is rational and $g(x) = 0$ whenever x is irrational is not Riemann integrable but has Lebesgue integral equal to 0 (since the set of rational numbers has measure 0).

The Lebesgue approach—that is, the approach based on measures, which generalize to much more complicated sets than $\mathbf{R}$—is also useful in probability theory. Indeed, calculating the probability of an event consists of "measuring" the set of elementary possibilities corresponding to this set. Thus a probability measure is just a special case of the measure defined above: more specifically, a probability measure is a measure p such that $p(\Omega) = 1$, where Ω is the set of all the elementary possibilities. (Since it is guaranteed that one of the possibilities will occur, the probability of the event Ω must be 1.)

But instead of choosing Lebesgue's approach, Bourbaki decided to develop a different one, which we will just briefly explain here. Bourbaki introduced integrals based on linear forms defined on the space of continuous functions. Such a linear form is a function L which associates to each continuous function f a number $L(f)$ such that $L(af_1 + bf_2) = aL(f_1) + bL(f_2)$ for any continuous function f_1 and f_2 and numbers a and b (this property, called linearity, is one of the key properties of integrals). This approach was criticized "because it uses topological concepts—like continuity—even though integration doesn't need topology," explains Jean-Pierre Kahane, a mathematician at the Université de Paris-Sud, adding that "Arnaud Denjoy, who was one of the eldest among the first Bourbakis and who invented an original, but quite complex theory of integration, was disgusted by Bourbaki's treatment of the subject." To be honest, however, it is important to emphasize that Bourbaki's approach did have some useful consequences. For example, it led Laurent Schwartz to create the theory of distributions, which generalizes the concept of a function and involves linear forms defined on a narrower class of functions than that of continuous functions.

Un exemple d'animosité envers Bourbaki est ce double poème d'un goût contestable écrit (en anglais) par Arnold Seiken, mathématicien aux États-Unis, à propos des mathématiques françaises :

Sublime

Pêle-mêle
Henri Poincaré
D'élégants théorèmes
Prouva sans façons,
Classa des variétés
Difféomorphiquement
Utilisant la dualité
Et du génie aussi.

Autre

Pêle-mêle
Nicolas Bourbaki
Ressuscita de parmi les morts,
Pour les mathématiques sauver,
Débita d'assommants traités
Hypercinétiquement,
Prouvant qu'il aurait dû
Dans la tombe rester.

The Mathematical Intelligencer, 1995

(For English translation, see appendix.)

Hyperaxiomatizers Lacking Generalization?

Let's return from these digressions to the criticisms made against the style (in the general sense of the term) of Bourbaki's treatise. One of the first things that struck readers was the formal and abstract nature of the text. The concision of the sentences, the abundance of mathematical symbols, and the extreme dearth of figures clash with the style of textbooks from the beginning of the century. People also accused Bourbaki of excessive formalism and generality. Today, such criticism seems unjustified. A quick visit to a mathematics library at any university would reveal many more formal books than Bourbaki's and books with a much higher number of symbols per square inch. As for generality, Armand Borel, a former member of Bourbaki, explains in an article published in 1998 that "Contrary to my first impressions [...] the book's goal is not to achieve the greatest generality possible. Instead, it is to achieve the most effective generality, the one most able to meet the needs of potential users in various areas. They left out any refinements of theorems that excited specialists without substantially extending the theorem's applicability. Of course, later developments could show that Bourbaki didn't always make the best choices. Nevertheless, this served as a guiding principle." Christian Houzel, who was never a member of Bourbaki, stated a similar opinion: "Bourbaki never tried to obtain maximal generality. On the contrary, the Bourbakis aimed for the minimal generality that could encompass what they knew. This is why the level of generality has evolved with the newer editions of the treatise."

The perception of excessive generality may stem from Bourbaki's style of explanation—progressing from general to specific—and it is certainly not improved by the impersonal tone of the treatise. This aspect of the treatise's style even attracted criticism from Alexandre Grothendieck, who was a "hyperaxiomatizer lacking generalization"—a nickname that, according to Pierre Samuel, the Bourbakis gave to any members who pushed too hard towards abstraction and generalization. In his lengthy *Récoltes et semailles* ("Harvests and Sowings"), which was never published commercially, Grothendieck writes (around 1985) that "the canonical text hardly gave an idea of the atmosphere in which it was written, to say the least. By now I've realized that this is in fact one of the main shortcomings of Bourbaki's writing—that not even an occasional smile is included to suggest that these books were written by people, and people connected by much more than some oath of unconditional fealty to the merciless canons of rigor..."

The criticisms of Gustave Choquet, who was never a member of Bourbaki but who contributed greatly to reforming mathematics education in the universities, are more severe. In an interview with Marian Schmidt published in 1990, he said that "most French mathematicians of my generation acquired much of their mathematical education from Bourbaki. Bourbaki influenced their style and their work and transmitted some of its qualities and flaws. What were these flaws? It seems that any group that works for a long time in isolation is condemned to dogmatism. I think this is the main criticism against Bourbaki: the basic definitions and theorems are presented without any justification or heuristic explanation. They have the desiccation and meagerness of bones whose meat, despite being delicious, has been thrown to the exercises. The reader who neglects to do them all ends up with only a caricature of mathematical activity."

Arnaud Denjoy (1884–1974) was disgusted by the way Bourbaki presented integration in its treatise.

Bourbaki Algebrized Analysis

Bourbaki's penchant for algebraic concepts and methods also has its drawbacks. The group's initial goal was to write a treatise on analysis; instead, the group achieved an algebrization of analysis, which was fruitful in many ways. As Laurent Schwartz writes, "The most clear benefit I think I got out of Bourbaki was to be 'algebrized.' I'm an analysist by nature, and all my work deals with analysis and probability. But I use algebra and algebraic methods as often as possible [...]. I am one of the most algebraic among the analysts. It's Bourbaki who gave me this." However, this algebraist attitude does not suit everyone, and especially not geometers. Marcel Berger, an eminent geometer, was a Bourbaki guinea pig but didn't get interested in the group. Bourbaki's spirit also failed to seduce the differential topologist René Thom, who believes in the importance of geometric intuition. Benoît Mandelbrot, a wholehearted geometer, is one of the most fervent rejecters of Bourbaki. The nephew of Szolem Mandelbrojt (one of the first participants in Bourbaki), Mandelbrot left l'École Normale Supérieure in favor of l'École Polytechnique to escape from the Bourbakis. "Thanks to my uncle, I realized this this group was a militant bunch, that they had strong prejudices against geometry and against every science, and that they were inclined to scorn, or even humiliate, anyone who didn't follow them," he said in an interview published in 1985. He even left France for the United States in 1958 because of the group's "stifling influence" and the predominance of its choices.

Gustave Choquet

But this era is no more. Bourbaki no longer reigns over French mathematics, and even less over world mathematics. Geometry has become fashionable once again, and not only in mathematics. In theoretical physics, for example, the geometry and topology of 10- or 11-dimensional spaces are studied with the goal of constructing a theory that encompasses all the forces and fundamental particles. Geometry has also entered computer science with image processing, realistic visualization of three-dimensional scenes, robotic vision, and so forth. The division between pure and applied mathematics is weakening: even number theory and algebraic geometry have applications in cryptography and information coding. The interactions between physicists and mathematicians have developed in abstract domains like the theory of elementary particles as well as in concrete problems like the turbulence of a flowing fluid. Today, many scientists and engineers study mathematics connected to computer computations, such as numerical calculation, formal calculation, and optimization algorithms. The

Laurent Schwartz at the Bourbaki conference in Pelvoux (Summer 1951).

mathematical atmosphere is no longer that of the 1960s, Bourbaki's golden age. The orderly ranks of mathematicians working "for the honor of the human spirit" have been replaced by plentiful mathematicians working in every direction. The heat raised by the Bourbakian enterprise has dissipated. And, now that this undivided reign of pure mathematics has ended, some are predicting a dictatorship of applied mathematics.

Claude Chevalley (1909–1984)

Claude Chevalley was born in Johannesburg, South Africa, on February 11, 1909. His father, a diplomat, was at the time a consul general. Chevalley's education started at the elementary school of Chançay, a small town near Tours, where his father had acquired a property in 1910. He continued his education at the Lycée Louis-le-Grand in Paris. Although he had been interested in science for a long time, Chevalley did not decide to pursue a career in mathematics until his last year of high school, when he was inspired by an excellent teacher, G. Dufour. In 1926, he entered l'École Normale Supérieure at the age of seventeen. One of his friends there was Jacques Herbrand, who influenced him greatly. (The brilliant Herbrand was interested in mathematical logic, which was an unexplored area at the time in France, and made important contributions to this area before his premature death at the age of twenty-three in a mountaineering accident in 1931.) At l'École Normale, Chevalley also met André Weil, who had just returned from his trip to Italy and Germany. It was with Weil that Chevalley was initiated to the modern aspects of algebraic number theory, which would become one of his main areas of interest. In the academic year 1929–30, a grant allowed Chevalley to carry out his first scientific research, following which he served in the military. From late 1931 until 1936, Claude Chevalley was supported by a grant from the Caisse Nationale des Sciences (the National Science Fund, a precursor of today's Centre National de la Recherche Scientifique) and devoted himself to mathematics, philosophy, and social and political criticism. He spent the academic year 1931–32 in Hamburg, Germany, where he studied under Emil Artin and completed his doctoral thesis. He then spent several months in 1933 in Marburg, where Helmut Hasse was teaching. He married his first cousin Jacqueline in November 1933.

In 1936, Chevalley taught for one semester at the University of Strasbourg as a substitute for Weil, who had been invited to visit the United States. He was an assistant professor at Rennes during the academic year 1937–38 before spending one year at the Institute for Advanced Study in Princeton. At the outbreak of World War II, the French ambassador advised Chevalley to stay in America. Following this advice, Chevalley worked at Princeton University until 1948, when he spent a year in Paris on a Guggenheim Fellowship before moving to Columbia University in New York. Meanwhile, he divorced his wife in 1948 and married Sylvie Bostsarron, a specialist in the history of theater. The couple had one daughter, Catherine Chevalley, who is now a philosopher at the University of Tours. Claude Chevalley returned to France in 1955 to accept an appointment at the Sorbonne, although some mathematicians thought this was an unjust reward for someone who had spent the terrible years under good conditions, and campaigned against his appointment. Chevalley remained at this university until his retirement in 1978. He died on June 28, 1984 after a long illness.

Chevalley's contributions to mathematics are numerous and important. Most are in algebraic number theory (particularly in class field theory), algebraic geometry, and group theory. He also wrote books on these areas, which have now become classics.

Philosophy is the other side of Chevalley's intellectual life. Like Herbrand, he combined rigor in mathematics with very personal experiences of anguish and liberty. He was interested in the epistemology of Émile Meyerson, one of his father's friends, who believed that science reflects the desire to decrease fundamental anguish by discovering stability and lasting truths. For Chevalley, as his daughter explains, mathematics can only be created independently of any exterior purpose. It cannot and should not submit to the reasoning required by, for example, the physicist's reality or the desire to uncover the underlying structure of the world. In other words, Chevalley held the Bourbakian view that one must eliminate reality to do pure mathematics. He was also influenced by another friend, the philosopher Arnaud Dandieu. Along with Robert Aron, Dandieu (who died young in 1933) was an important figure in the Ordre Nouveau movement created in 1930. This was a group of intellectuals of anarchist and European nature who called themselves "personalists" and advocated a new order based on human primacy, direct democracy, an anti-productivist economy, and federalism. Swept into this movement (which has nothing to do with the group of extreme rightists who later took the same name) by Dandieu, Chevalley became one of its key figures until the movement dissolved in 1938. Chevalley concentrated more on mathematics after World War II, but during the sixties he campaigned in a group of intellectual ecologists (with two other Bourbakis, Alexandre Grothendieck and Roger Godement), which published a journal called *Survivre et vivre* ("Survive and Live"). According to those close to him, Chevalley always remained true to the convictions of his youth, and especially to his quest for rigor and liberty. Among the founders of Bourbaki, he, without a doubt, was the most individualistic and the one who best kept a critical eye on their endevour.

10

New Math in the Classroom

In the 1970s the wave of New Math surged into secondary education. Structures inspired by Bourbaki were certainly included, but how much was the group itself involved?

"Definition 4: A subset I of an ordered set E is called an *interval* of E if it satisfies the implication $(x \in I \text{ and } y \in I \text{ and } x \leq z \leq y) \Rightarrow (z \in I)$."

"Theorem 6: The identity is the only automorphism of the field **R**."

"Theorem: Addition and composition of functions equips the set $L(V)$ of endomorphisms of a vector space B with the structure of a unitary ring."

"Every involutive endomorphism is a symmetry."

"Definition: We say that an orthogonal endomorphism of ϕ of $\mathbf{E}_3$ is a vector rotation to express the fact that the subspace of variables invariant under ϕ has dimension 1 or 3."

"Theorem 2: Let F be a vector function and x_0 be an accumulation point in its domain of definition. Then $\lim_{x \to x_0} \overrightarrow{F(x)} = \overrightarrow{0} \Leftrightarrow \lim_{x \to x_0} \|\overrightarrow{F(x)}\| = 0$."

THEOREM AND DEFINITION

Given: a graduated line (Δ, g).

1. For each pair (a', b') of real numbers such that $a' \neq 0$, the function g' from Δ to **R** defined for any element M of Δ by $g'(M) = a' \cdot g(M) + b'$ is bijective.
2. The family of all bijections thus defined has the following property:
   For any two bijections g' and g'' of this family, there exists a pair (a, b) of real numbers such that $a \neq 0$ and $g''(M) = a' \cdot g'(M) + b$ for every element M of Δ.

Any bijection belonging to this family is called a graduation of Δ. The number $g'(M)$ is called the abscissa of M in the graduation g'.

A property of a graduated line and the definition of a graduation...in eighth grade!

Laurent Schwartz teaching about the theory of distributions. Laurent Schwartz's course revolutionized mathematics teaching at l'École Polytechnique.

Are these statements from Bourbaki's *Éléments de mathématique*? No. Or maybe from mathematical courses for university students or for students preparing to enter the presitigious Grandes Écoles? Once again, no. In fact, they are simply extracts of a French mathematics textbook for high school seniors, published in the 1970s by Hachette in their Aleph$_0$ series. And this textbook consists of six volumes—two for algebra, two for geometry, two for analysis—for a total of over 1500 pages! Its content would leave today's math student stunned.

In France and around the world, this was the era of New Math, an ambitious reform in elementary and secondary math education. However, it was a failed reform whose biggest achievements were wasting gallons of ink and creating heated controversy among parents, teachers, mathematicians, educational psychologists, and others. It was a setback whose results remain today as scars in the educational system. Some people connect Bourbaki with this reform. Why? Are they right?

Bourbaki Attacks Higher Education

Before tackling the subject of seconday education, let's discuss the reform of higher education. At least in France, Bourbaki clearly played an active role in this reform. Indeed, this was the very nature of the group's initial project, the project to write a new treatise on analysis that would replace the obsolete and unsuitable texts used at the time in higher math education. The Bourbaki pioneers, who for the most part were professors at universities outside Paris, started to modernize math education in their respective institutions as soon as the group had been formed. For example, Henri Cartan said that "at Clermont-Ferrand, where the University of Strasbourg had moved in 1939, I taught differential calculus on Banach spaces. This was a real revolution." In this way, students at various universities outside the capital were introduced to slightly more modern mathematics than what had been taught to students just a couple years before. This revolution arrived in Paris somewhat later. While the students of l'École Normale Supérieure didn't have to wait long to benefit from Henri Cartan, who arrived in 1940, students at other Parisian institutions had to wait until the arrival of Gustave Choquet in 1954. According to Henri Cartan, at the end of each academic year the dean would hold a meeting with the math faculty to distribute the teaching of the math classes for the following year. The class on differential and integral calculus had been taught for many years by Georges Valiron, who gave traditional and boring lectures. One year, though, illness prevented Valiron from continuing to teach. At this point, Henri Cartan recalled, "I recommended Choquet, who shared our point of view on teaching." So, Gustave Choquet, who was never a member of Bourbaki, took on the job. As Choquet described to Marian Schmidt, "Until then, the course mostly followed the elementary sections of Goursat's treatise. I modified the focus and content of the course. I thus had the good fortune of being behind the revolution in French mathematical teaching, at first in the classes for advanced university students, and later, as the reforms spread contagiously, for the younger students as well." However, this revolution was not accepted by all, and conservative mathematicians protested bitterly. "For example, I remember that Henri Villat demanded, 'How can you expect students to understand when I don't understand it myself?'" Henri Cartan recalled.

In his autobiographical book *Mathématique: (récit)* (which mentions Bourbaki several times), the author-mathematician Jacques Roubaud describes how students received Choquet's revolution: "Faced with the sudden metamorphosis of mathematics occurring as they watched (and especially as they listened), even the most hardened students, those who had taken classes preparing them for the Grandes Écoles entrance exams and who had survived the slaughter of the exams to earn their certificate in general mathematics, felt their most established convictions waver. Over the course of their studies, these students had developed a stable and familiar view of mathematics, and now this view

Henri Cartan in Bombay, 1960.

was changing so much that it was becoming locked away, sealed before their eyes. And they didn't find the new face of mathematics to be especially pretty. Those students who had had to repeat a year were the most noticeably distraught, as they found nothing in common between Choquet's classes and Valiron's of the year before. It was as if this science had been replaced by a new one over the summer vacation, a new science that had kept the name of the old one merely out of convenience." Choquet quickly gained reinforcements. By the following year, several professor and lecturer positions had been vacated. According to Martin Andler, a "fierce battle" ended with victory on the part of the revolutionaries. The new faculty members for the 1955 academic year were Chevalley, Ehresmann, Pisot, Zamansky, Godement. All of these, except Zamansky, were at some point members of Bourbaki.

Bourbaki Reaches l'École Polytechnique

The situation at l'École Polytechnique also deserves to be discussed. From World War I until the 1950s, teaching at this elite school—including the teaching of mathematics—was mediocre, despite the presence of Paul Lévy (a great probabilist whose work was long unrecognized) and of Gaston Julia, who at the time was nearing the end of his career. The hiring of Laurent Schwartz (a member of Bourbaki) in 1959, following the retirement of Paul Lévy (Schwartz's father-in-law) gave new life to this institution, which had produced almost no new mathematicians for many years. Schwartz's course on analysis was very succesful, and the changes he introduced—despite resitstance—led to a profound revival of teaching at l'École Polytechnique, especially after 1968. Jean-Pierre Bourguignon described how "there was such a contrast between Schwartz's teaching and the teaching of other subjects that the need to completely renovate the other courses became obvious." According to Bourguignon, the success of this renovation was so great that l'École Polytechnique today produces mathematicians of the highest quality: "At the most recent international congress of mathematicians, the one in Berlin, four of the ten plenary addresses were given by French researchers, and three of those four had graduated from l'École Polytechnique."

Although Bourbaki actively contributed to the renovation of higher mathematics education, this renovation was not entirely the work of Bourbaki. For one thing, this contribution is not the work of Bourbaki (excepting the indirect influence of *Éléments de mathématique*) but rather the work of individual members. It is highly unlikely that

Jean Piaget in 1977.

Bourbaki developed a collective strategy to obtain key positions at universities or the Grandes Écoles (including l'École Normale Supérieure and l'École Polytechnique) or to redefine the content of mathematics courses at these institutions. In addition, the members of Bourbaki were not the only mathematicians to renovate higher mathematics education. Gustave Choquet is the most obvious non-Bourbaki to contribute to this revolution, and others include Jean Leray and André Lichnerowicz, who were appointed as professors at the prestigious Collége de France in 1947 and 1952 respectively.

It is much harder to assess the role of Bourbaki in the New Math movement in secondary education. This famous reform started in the 1950s, when it became apparent that the mathematics taught in secondary schools was worn out and poorly adapted to the modern demands of economics, technology, science and culture. This realization affected many countries, both in France and around the world. But what exactly triggered it?

In the 1950s and 1960s, the West was undergoing a profound transformation that involved economic and industrial growth, scientific and technological progress, and cultural change. Training excellent scientists and engineers was considered necessary for economic development. Furthermore, the competition between the Soviet Union and the West, intensified by the launch of the first Sputnik in 1957, made the West worried about falling behind in technology. A solid knowledge of mathematics was considered necessary for all scientists and engineers.

The evolution of mathematics itself added to this situation. Partly due to the influence of Bourbaki, mathematics had become seen throughout the world as a unified subject based on set theory and built out of general structures (groups, rings, fields, etc.) defined by axioms. This sort of mathematics was only vaguely related to that being taught in elementary and secondary schools, despite the mathematical revolution taking place in universities. Thus mathematicians, as well as young teachers who had taken Bourbaki-influenced classes in college, wished to modernize the teaching of their subject.

André Lichnerowicz (1915–1998) in 1988.

Mathematics Was Everywhere

Linked to this evolution was the idea that mathematics (or the singular noun *mathématique*, as Bourbaki called the subject) was a universal language intended to be used in all disciplines, and in particular in all sciences, including even humanities and social sciences. This went with the modern trend in mathematics of emphasizing the relations

At the 1954 Bourbaki conference in Murols: Roger Godement, Jean Dieudonné, André Weil, Saunders Mac Lane, and Jean-Pierre Serre (from left to right). Roger Godement ardently supported the implementation of Bourbaki's philosophy in university-level math education.

between objects (numbers, functions, geometric figures, and so forth) rather than the objects themselves, whose characteristics were considered to be of little importance. Many declared that "mathematics is everywhere," meaning that it is an essential part of everyone's academic and cultural knowledge. It is likely that this point of view was based partially on the fashion of structuralism affecting philosophy, literature, ethnology, linguistics, and psychology.

A final factor in sparking the New Math revolution was the new trends in pedagogy. Jean Piaget saw an analogy between the mental structures created when a child learns mathematics and the mother-structures (algebraic structures, ordered structures, and topological structures) that Nicolas Bourbaki discussed in its article *L'architecture des mathématiques* ("The Structure of Mathematics"). Furthermore, Piaget and many other educational psychologists emphasized the importance of active learning in a child's intellectual development, which created a movement for teaching based on observations, experiments, and inferences made by the students with the help of their teachers rather than on knowledge obtained directly from the teacher. The Bourbaki style of mathematics seemed better suited to this new sort of teaching than the traditional style did. In addition, mathematics seemed to be more democratic than other subjects, since its emphasis on concepts required no cultural prerequisites. Many argued, erroneously or otherwise, that students would find math more accessible than Latin and Greek, which the schools traditionally used to choose their top students. Such an argument was by no means insignificant in a time when more students were continuing their education through high school and in the social and political atmosphere that led to the May 1968 student uprisings in France.

"Down With Euclid!"

For most countries, the New Math revolution consisted of four phases. The first was a phase of realization and reflection. An important event of this stage was the November 1959 Colloque de Royaumont, a ten-day conference organized by the Organisation for European Economic Cooperation (now called the Organisation for Economic Cooperation and Development) to promote a reform in the content and methodology of secondary math education. It was during this conference that Jean Dieudonné, one of the participants, threw out a provocative declaration in reference to the teaching of geometry: "Down with Euclid!"

The years from 1964 through 1967 marked the second phase, a preparatory one in which committees formed. The third phase saw educational experiments carried out and new curriculums implemented. The fourth and final stage, which occurred during the early 1970s, was the general implementation of the new curriculums. In France, one important step in the New Math reform was the government's creation of the Lichnerowicz Commission in 1967. Chaired by André Lichnerowicz, this committee comprised a total of eighteen university and high school teachers, including Gustave Choquet, the physicist Louis

2.1 CORPS $\mathbb{C}$ DES MATRICES $\begin{pmatrix} a & -b \\ b & a \end{pmatrix}$

2.1.1 Définition.

On appelle $\mathcal{M}$ l'ensemble des matrices carrées d'ordre 2 à termes réels. Sur le corps $\mathbb{R}$ des nombres réels, on appelle $\mathbb{C}$ le sous-ensemble des matrices $M(a, b)$ de la forme :

$$M(a, b) = \begin{pmatrix} a & -b \\ b & a \end{pmatrix},$$

pour tout a et pour tout b.

The definition of complex numbers in terms of matrices, as taught to twelfth-grade students.
(For English translation, see appendix.)

Néel, Pierre Samuel (a member of Bourbaki at the time), Charles Pisot (a former member of Bourbaki), and André Revuz (one of the main proponents of the reform). As the job of the Lichnerowicz Commission was to propose new curriculums, this committee played a key part in the New Math reform.

It would be tedious to describe the new curricula in detail, but their general principles are simple to explain. The new curricula introduced the principles of mathematical and formal logic and of naive set theory earlier than the old curricula, including an introduction to groups, rings, fields, and vector spaces presented axiomatically; introduced complex numbers (in senior year) and principles of probability theory; and replaced traditional geometry with linear algebra (which deals with linear equations and vector spaces). The new curricula emphasized rigor in definitions, theorems, and proof-writing while deemphasizing all numerical, algebraic, and trigonometric calculations. The reforms in other countries were of the same nature as those in France.

A Revolution and a Counterrevolution

The reform certainly went too far, especially in the implementation of the curricula in classrooms and textbooks. The results of the reform quickly aroused harsh criticism, which the media reported to the general public. The abandonment of geometry, or more precisely its algebrization, was one of the main targets of the complaints. Another was the excessive formalism and abstraction of New Math. "[...] New Math is not in touch with reality," and "[...] Set theoretical algebra does not teach students how to reason" are just two of the many criticisms published by the magazine *Sciences et Vie* in a series of articles (1971–1972) attacking the reform. Eminent mathematicians like Jean Leray and René Thom also charged against New Math. As Leray wrote

THEOREM AND DEFINITION

For any two pairs $(D_1, D_1{}')$ and $(D_2, D_2{}')$ of vector rays in E_2 the relation "There exists a vector rotation f of E_2 such that

$$f(D_1) = D_1{}' \qquad \text{and} \qquad f(D_2) = D_2{}'\text{"}$$

is an equivalence relation in DxD, where D is the set of vector rays in E_2.

An equivalence class for this relation is called **an angle of two vector rays in E_2.**

The definition of the angle between two rays, as taught in 1971 to eleventh-grade students...with no diagram!

in the October 1971 issue of *Gazette des mathématiens*, "'New Math' is a series of concepts that are defined without reference to their characteristic properties (axioms) and that fail to lead to statements of any remarkable properties. It's impossible to reason with these concepts or to find interest in them. Learning them is a test of memory that poisons intelligence." Jean Dieudonné denounced it as "a new method of education, a more agressive and stupid method that flies the banner of modernism."

The counterrevolution was in vain, and the New Math revolution continued to be criticized around the world as a global failure, even while this failure was never measured quantitatively by statistics. "There's no doubt about the failure. The teachers complained of being unable to teach according to the curricula and textbooks prescribed by the ministers, the students didn't seem to understand these 'unifying concepts,' and the parents realized that their children could no longer count or solve problems," explained Anna Sierpinska, a Polish woman living in Canada who has served as the vice-president of the International Committe on Mathematical Education. But did New Math fail? To what extent was the new material understood less than the old material? The new curricula did have a certain coherence, so why did their implementation create such extreme problems? An objective and detailed analysis of these questions is lacking even to this day. Some hints at answers do exist in criticisms expressed here and there, but they don't create a clear and global view of the reform's problems. The reasons for the failure remain poorly understood, concealing any lessons that might be learned for the future.

Gustave Choquet and his revolutionary booklet.

LES COURS DE SORBONNE

Certificat d'Etudes Supérieures
de CALCUL DIFFÉRENTIEL et INTÉGRAL

ALGÈBRE DES ENSEMBLES
ALGÈBRE

PAR

M. CHOQUET

CENTRE DE DOCUMENTATION UNIVERSITAIRE
5, PLACE DE LA SORBONNE, PARIS V

On the other hand, we do have the retrospective comments of Laurent Schwartz and Gustave Choquet, two mathematicians who were also renowned as teachers. As Choquet told Marian Schmidt in 1990, "in an abstract and sterilized world, the progress of a hundred years would allow us to elegantly and rigorously present fundamental concepts and theorems *ab ovo*, without referring to experience or geometric intuition. This is what happened, both in France and in other countries, with the drama of the mathematics reform at the end of the 1960s. Jean Dieudonné's famous proclamation 'Down with Euclid!' represents the focus of the ministerial committee in charge of developing new curricula for mathematics education in middle schools and high schools. The motivating idea of the movement was that, since fundamental concepts are necessary for all logical constructions, these concepts (including logic, set theory, algebra, and linear algebra) should be taught first. The result was bound to be catastrophic,

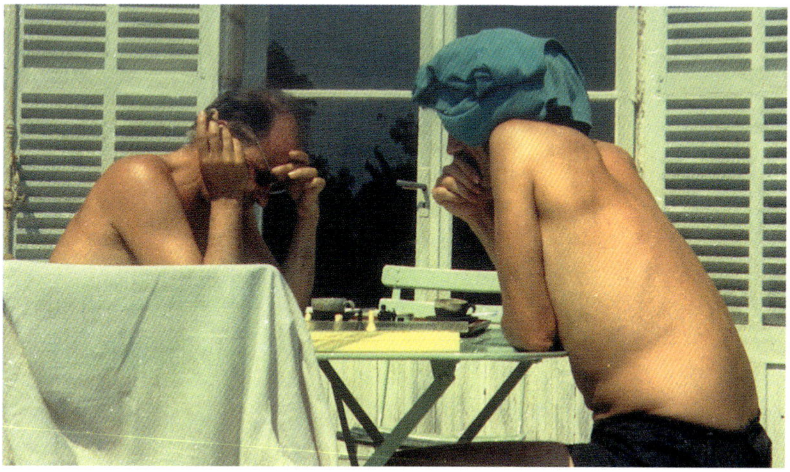

François Bruhat and Michel Demazure in la Messuguière (near Grasse), July 1975.

since all pedagogical considerations, such as the students' motivation and previous knowledge, the teachers' education, and the writing of reasonable textbooks, were swept aside. Also, little effort was made to meet the demands of physicists and engineers." In the official report *La France en mai 1981: l'enseignement et le développement scientifique* ("France in May 1981: Scientific education and development") published in December 1981, Schwartz writes that "[...] the teachers, parents, and children didn't learn modern mathematics but rather just the basic language of an extraordinarily vast and modern subject [...] of which the definitions given in classrooms (around the world!) were only the ABCs of the subject. [...] Little by little, all the richness of the mathematics previously taught in high schools, all the theorems, geometric figures, and connections with other sciences, has been replaced by a plethora of axioms and definitions. Most students find these incomprehensible, and the effect is that the results are *very poor*. Mathematics is rich when it introduces few concepts and structures and many theorems, but the New Math taught in schools introduces an enormous number of concepts and definitions and almost no theorems. This is *very poor* mathematics. [...] The goal of mathematics is *not* to prove rigorously things that everyone knows. Instead, the goal is to find rich results and then, in order to make sure they are true, to prove them."

So what role did Bourbaki play in the introduction of New Math? Dieudonné expressed his personal opinions on education with provocative exclamations, and while he didn't play an official role in developing the new curricula, he did greatly influence them (which is apparent, for example, in the algebraic presentation of geometry). Pierre Samuel

Jean-Pierre Kahane believes that the Sierpinski triangle, a fractal structure, is an analogy for the way mathematics curriculums are being drained of their content more and more each year.

participated in the Lichnerowicz Commisision, but so did seventeen other people. In addition, Samuel was not among the most radical. "Lichnerowicz and I were quite moderate, but some of the committee members took things too far," he explained. Cartan and Schwartz gave addresses on contemporary mathematics to the *Association des professeurs de mathématiques*, an association whose members included many high school teachers enthusiastic about the reform. This was basically the extent of the individual contributions by Bourbaki members. As a group, Bourbaki took part neither in the reform nor in the debates surrounding it. According to Pierre Samuel, "Bourbaki held no opinions on high school education. The group didn't even hold any opinions about education in the first years of university, although some members wanted to write a mini-Bourbaki for these earlier years. This idea was abandoned, as we already had a lot to do and there were already good books for this level."

Supicious but Silent

Michel Demazure recalls that the group viewed this reform with much suspicion, and that certain members were strongly against it: "What we all shared was contempt for the pedagogical movement. We were preoccupied with the content of courses, not how to teach them." But although Bourbaki did not participate in the reform, the group did influence it indirectly. According to Anna Sierpinska, "Bourbaki's influence on this movement was particularly evident in the underlying mathematical philosophy of the choices and in the organization of the mathematical content of the new curricula. The goal was to build mathematical knowledge from the earliest classes, starting even in preschool, as an enormous, unified structure built on general concepts like sets, orders, relations, and groups." Thus Bourbaki's vision of mathematics had spread to the world of mathematicians and then to that of higher education; from there, it spread to high school teachers who proposed using it to revolutionize secondary math education, sometimes even explicitly mentioning Bourbaki in their arguments. But Bourbaki, who never claimed that the methods used in the group's treatise could be displaced to secondary education, didn't hold itself responsible for the misinterpretations of its philosophy and didn't try to perfect the extension of its philosophy to high school teaching. This attitude may deserve to be criticized, and indeed Pierre Cartier did: "Bourbaki acted very hypocritically in terms of the reforms in secondary education. The group influenced the movement quite a bit, but it denied any responsibility for the results." Meanwhile, Michel Demazure

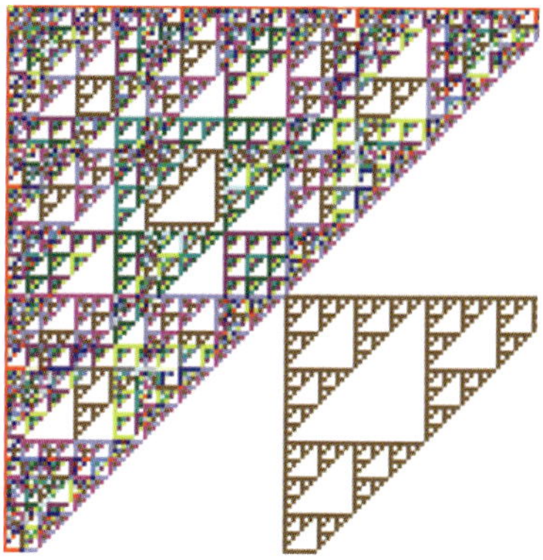

saw an element of Jansenism in the attitude held by Bourbaki, whose members were mostly Protestant: "Bourbaki didn't try to justify itself, and the group didn't feel responsible for the misinterpretations others made of its work. The Bourbakis were all against the use of the treatise as a model for teaching, but they held a certain Jansenist philosophy that said, 'Let people believe what they will."'

Whatever the cause and extent of their failure, the various reforms of the New Math revolution were abandoned towards the end of the 1970s. Following New Math was a counterreform, leading to much less ambitious curricula. Traditional geometry was reintroduced, superfluous formalism was abandoned, calculations were reemphasized. But today's curriculums are also questioned. They seem less coherent, not as well constructed. Each year, more and more topics are removed to make the curricula easier, to the point where Jean-Pierre Kahane, a mathematician who chairs a committee on mathematical education, said that "today's curricula are like Sierpinski's triangle." Vector spaces are no longer included, even though they don't seem to pose any exceptional difficulties and play a considerable role in mathematics and its applications. In addition, number theory has almost completely disappeared. Today's curricula demands little from students' imagination and reasoning abilities and fails to prepare students to create and write proofs. "Students are not really taught to reason," Pierre Samuel complains, while Michel Demazure laments that "now, most of the exercises almost contain their answers in their statements." It seems that abandoning the New Math revolution was throwing the baby out with the bathwater. Today, mathematics inspires in many students as much fear, lack of interest, and even hatred as ever. And what Dieudonné wrote in his 1987 book *Pour l'honneur de l'esprit humain* unfortunately remains true today. "Nothing taught in high school mathematics was discovered after 1800," he writes, although it would be more accurate to replace "nothing" by "almost nothing."

11

An Immortal Mathematician?

Nicolas Bourbaki continually replaces its members, yet the group has not been able to maintain its youth. Seventy years after Bourbaki's creation, its role and survival are uncertain.

"Forty years of Bourbaki: The famous mathematician is immortal but aged," states the April 9, 1980 issue of *Le Monde*, while the April 28, 1998 issue of *Le Monde* announces that "Bourbaki is dead, QED." As the media rarely reports on mathematics and the lives of its servants, the publication of these articles by important newspapers adds weight to the question about Bourbaki's survival. Now, at the dawn of the twenty-first century, Bourbaki still exists but the group is no longer what it was in the fifties and sixties. The volumes of *Éléments de mathématique* are no longer published at the same rate as before, with the most recent volume published in 1998 and the one before that in 1983. Meanwhile, the content of the older volumes is starting to become dated. The publishing house Masson maintains that no new volumes or new editions are in sight, stating that "the publication of Bourbaki's treatise is dead, we're not planning to develop it any further." And the mathematical community no longer speaks of Bourbaki, neither of the group's pranks (to the extent that there still are any), nor of its mathematical choices. The seminars alone fly the banner of Bourbaki and continue to attract visitors, three times a year.

The Mathematical Atmosphere Changed

So what happened? Why was Bourbaki already considered old in 1980, even before the group's fiftieth birthday? Shouldn't the constant replacement of its members guarantee the group eternal youth? A whole

The April 26, 1998 issue of *Libération* and the April 9, 1980 issue of *Le Monde*. The headlines declare, "Bourbaki is dead, QED" and "The famous mathematician is immortal but aged."

series of factors led to Bourbaki's decline, some dating back to the 1950s. It was during this decade that the first part of *Éléments de mathématique*, the first six books dealing with the fundamental concepts essential to all mathematics, was published. As Armand Borel wrote in 1998, "[...] mathematics had developed enormously, the mathematical atmosphere had changed considerably, partly thanks to Bourbaki's work, and it became clear that we couldn't continue following our old plan. Although it wasn't intentional, the founding members often had more influence in important decisions, but they were starting to retire and most of the responsibility was being handed down to the younger members. Some of the basic principles needed to be reexamined." One such principle was the strict linear order that the chapters and books of the treatise were supposed to follow, since following this principle would considerably delay the publication of certain volumes.

Furthermore, Borel explained, Bourbaki had become popular and other books with similar styles were being published. Ignoring these other books would lead to double work and wasted time, while taking them into account would perilize the autonomy of Bourbaki's treatise. Another Bourbaki tradition that fell under question was the involvement of every member in the writing of every chapter. While this noble principle was relatively easy to implement for the volumes dealing with the fundamental subjects, it was much harder to continue

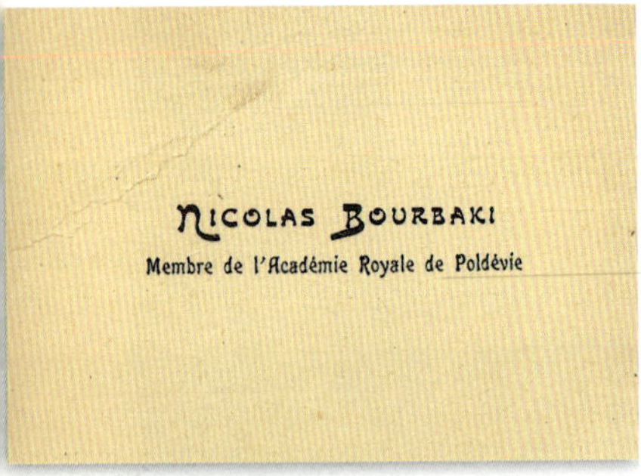

To establish Nicolas Bourbaki's authenticity, some members carried his business cards.

it for the more specialized chapters; however, Bourbaki balked at the idea of creating subgroups to write these chapters. Overall, according to Borel, the group was faced with two possible moves. One possibility was to cover all the fundamental areas of mathematics in a self-contained treatise, while the other, more pragmatic, possibility was to focus on specific topics that the group could handle without covering all the fundamental concepts in optimal generality. Bourbaki took some time to decide, but in the end the group chose to follow the second approach. This was a turning point in the history of Bourbaki: the group set off along a route that could be described as less messianic, insofar as many of Bourbaki's objectives had been achieved with the publication of the treatise's first six books.

This turning point coincided with the retirement of Bourbaki's historic core, which affected the atmosphere reigning over the group. According to Pierre Cartier, "the noise level decreased after Weil and Dieudonné left," which affected the violent atmosphere of the discussions, and "the relations among members became less tense, we had an atmosphere of much greater serenity." But Weil was also the member who dominated intellectually, the one who had wanted to create the group in the first place and had created the rules for how the group would work, and so his absence was bound to leave a gap among the remaining members. Dieudonné's presence was missed as well, although for a different reason. As Pierre Cartier described, "we realized that without his incredible ability to work, we were unable to keep working at the same pace."

Lack of Time or Lack of Enthuisiasm?

Many stress that today's members of Bourbaki don't have much time to devote to the group. "Bourbaki's main problem is time. In the beginning, Bourbaki could hold two-week-long conferences, but now that's inconceivable," said Arnaud Beauville, who retired from the group in 1997. According to Demazure, "Bourbaki demands a tremendous amount of time, the group needs to meet at least two months a year to produce real results." Hélène Nocton, who was the secretary of Bourbaki from 1959 through 1978, also believes that a mathematician's schedule is much fuller today. "One must attend many meetings, participate in committees, write reports, and so forth. Considerations about one's career are much more important than before, and, since participation is anonymous, a member of Bourbaki gains no direct benefit towards his career," Nocton explains. Yet she also maintains that in the late seventies Bourbaki met great setbacks from which it could not completely recover. Beauville partly agrees with this in saying that "there were certainly some unfortunate recruitments, some phantom members who accepted to be part of Bourbaki but who failed to come to the group's meetings. This disrupted the group's work." Cartier bitterly adds that "some people were recruited but never came. Some recent conferences lasted only two or three days and were attended by only three members! Currently, it's completely dead, empty, without projects, without plans to continue writing. Just take a look at the minutes in *La Tribu*. One time there wasn't even a report to read at the meeting!"

Grothendieck with his baby; as for his "Bourbakian baby," he rejected it.

According to some, another sociological element of Bourbaki's character also contributed to the project's decline: members were being hired for powerful positions and becoming more middle-class. When they created the group, the Bourbaki founders were nonconformist young Turks who wanted to reverse the established hierarchy in mathematics. But by gaining fame and influence, Bourbaki moved to the other side of the barrier. By the uprisings of May 1968 in France, Bourbaki had already become an establishment of mathematics. Alexandre Grothendieck describes this in his autobiography *Récoltes et semailles* as he summarizes the negative results of Bourbaki's move into the mainstream of mathematics: "It was certainly during the sixties that Bourbaki's 'tone' slipped to increasingly prominent elitism, a change in which I surely took part. [...] I still remember my astonishment when I discovered in 1970 the extent to which Bourbaki was unpopular among the lower strata [...] of the mathematical world. The name had become more or less a synonym for elitism, for strict dogmatism, for a cult that favored 'canonicalism' at the expense of living comprehension, for hermetism, for castrating antispontaneity, and that's not even all! [...] This group of exceptional quality exists no more. I don't know when it died, for it surely died without anyone noticing and sounding its death knell, not even within his heart of hearts. I suppose that an imperceptible degradation occurred in the members—everyone must have 'got on in years,' must have gone stale. They became important, prominent, powerful, feared, sought-after. Perhaps the spark remains, but innocence was lost along the way. [...] And respect was also lost along the way. When we had students, perhaps it was too late for the best to be passed on—there was still the spark, but there was no longer innocence nor respect, except for 'one's peers' and 'one's people.'"

In 1983, Pierre Cartier reached his fiftieth birthday and was retiring from the group, as Bourbaki's rules require. He had decided that the group's situation was quite negative. Bourbaki was finishing a long trial against its publisher Hermann, and Cartier believed that the trial's results were pathetic, that it "only made the lawyers richer."

Furthermore, the group's strength was waning, recovery seemed unlikely, and there was no longer any clear goal now that the group had more or less accomplished its initial mission. At one of the group's meeting, Cartier left the room to let his fellow members read his retirement letter. In it, Cartier presented his analysis of the situation and proposed that Bourbaki would fix a date, about two or three years in the future, before which all the current drafts would be completed with the help of everyone, including former members. "I proposed that in 1989, the fiftieth anniversary of the publication of the treatise's first volume, the group would dissolve and a deluxe edition of the entire treatise would be published. This would allow the initial enterprise to end with panache. And then we would be free to think about creating something else, something with a new goal but possibly the same people." But Bourbaki did not adopt Cartier's proposal. "It was a bit overdone," remarked Arnaud Beauville, and Cartier was accused of taking all the credit, of trying to say that Bourbaki would have no reason to exist without him.

"Its work is done and done well..."

Yet Cartier is not the only one to think that Bourbaki's days are over. Many mathematicians outside the group share this point of view, and sometimes expressed it in rather brutal terms. As one said, "Bourbaki is beyond comatose." Even former Bourbaki members hold similar views. In his autobiography, Laurent Schwartz writes that "[...] I think that Bourbaki's work is not as useful today. [...] The entire world has adopted Bourbaki's general ideas about definitions, concepts, reasoning, and written proofs, but its work is done and done well. The books are read by a fantastic number of students, but the group doesn't have many reasons left for existing. A current book by Bourbaki isn't of any greater importance than a book by any good mathematician outside the group." Henri Cartan agreed with this in an interview published in the August 1999 issue of *Notices of the AMS*: "What Bourbaki needed to do is done. Bourbaki won't live forever." Pierre Samuel says that "I get the impression that there is no longer the enthusiasm of the group's early years, that now whatever they can do is less new. I think Bourbaki should dissolve, but only after all the drafts in progress are completed. It would be stupid to close up shop before then."

In any case, most mathematicians agree that the time is no longer ripe for an enterprise like Bourbaki's. Mathematics has expanded considerably in innumerable directions. The number of researchers in the world has multiplied, as has the number of articles published. In the fifties, about 3000 articles per year were published, whereas today that number exceeds 100,000! No group of Bourbaki's size would be able to embrace everything, let alone deal with it in a unified way. As Christian Houzel explains, "Mathematics is moving too quickly; moreover, either there is no longer a need to revise mathematics to show the subject's unity, or such an endeavour would require a titanic effort."

This explosion of mathematics corresponds to a greater diversity in the intellectual approaches taken by researchers. The Bourbakian model, with its emphasis on axioms and structures, can no longer reign as king. The introduction of computers and computer science has made it particularly clear that Bourbaki's approach applies poorly to several entire branches of mathematics. "Bourbakian mathematics doesn't describe processes; apparently, it is impossible to find a general framework that includes everything about algorithms and programs," Michel Demazure explains. "No one has found a framework likely to apply to every case. There are some heuristic methods that cannot be axiomatized, methods for which no formal framework can be found. Bourbakian mathematics is somewhat analogous to Western musical notation, which is well-suited for writing down Bach, but which isn't at all suitable for jazz, for example. If one were to write a new treatise on mathematics, he would need to proceed in a more operative manner, emphasizing proofs by programs." According to Demazure, Bourbaki fell victim to the fundamental illusion that for any mathematical question there was exactly one right point of view to use, whereas in fact several angles of attack can be possible for the same topic.

General Bourbaki at the time of his surrender.

Mathematics is Unified by Its Branches

As mathematics has proliferated, the Bourbakian vision of this science has lost its pertinence. According to Jean-Pierre Kahane, "The unity of mathematics is not based on a single root, on set theory, as Bourbaki advocated, but rather on the fact that the various branches of mathematics communicate among each other." In other words, Bourbaki saw mathematics as a tree whose roots were set theory; however, using another botanical metaphor, it would be more correct to compare mathematics to a mycelium, a decentralized network of branched filaments.

The very style of mathematics practiced by today's researchers has moved away from Bourbaki's philosophy. While the height of the group's activity was during the supremacy of abstract algebra, recent years have seen a return towards geometry and concreteness (although of course high level mathematics is still very abstract!), especially in the interaction between mathematics and other disciplines. There is more pragmatism and less ideology. Moreover, the rise of applications of mathematics has led some researchers, especially Americans, to advocate mathematics that is not strictly rigorous. For example, some would like the statement of a theorem that is true with a probability of 90% to be a legitimate statement, as long as that probability has been evaluated using precise methods. Such proposals have recently created lively and interesting debates in mathematical journals. Although only advocated by a vast minority, these ideas show the existence of mathematical styles and approaches diametrically opposed to those advocated by Bourbaki.

Will Bourbaki survive its own decline and the incredible worldwide growth of mathematical production? Considering the current state of affairs, nothing is more uncertain. Yet its results remain, and its results are impressive. Materially, they correspond to a reference book of several thousands of pages and seminars whose notes, also several thousands of pages long, constitute a true encyclopedia that documents living mathematics. Intellectually, Bourbaki profoundly modernized mathematics and clarified its language and concepts. "One could compare Bourbaki's classification of mathematics with the immense revolution introduced by Linnaeus with his *Systema naturae* in 1758, in which he classified living beings, both animal and vegetable," Laurent Schwartz writes. Bourbaki's choices guided French mathematics and made France the world leader in algebraic geometry for over two decades. Sociologically, Bourbaki energized the mathematical atmosphere of its country, helping France earn back its place among the mathematics capitals of the world.

Finally, the group's alchemy helped many of the Bourbakis to become brilliant mathematicians, and the former Bourbakis don't hesitate to recognize this influence. As Henri Cartan writes in the preface to his *Collected Works*, "Working together with men of very diverse characters and strong personalities, united by a common drive for perfection, taught me very much. I am indebted to these friends for much of my mathematical knowledge." Similarly, Jean Dieudonné explains in his *Notice sur les travaux scientifiques* ("Notes on scientific work"), which he wrote for the Académie des Sciences in 1968, that "for many years during my career, I had the good luck to participate in a collective project to write a treatise covering many different areas of mathematics. This made me interested in very diverse problems, and undoubtably the stimulating exchanges of viewpoints this project created were the sources of many of my ideas for research. Although I can't reconstruct the exact origins of each of these ideas, I'd like to acknowledge my debt to the friends and colleagues who helped me in

General Bourbaki's unfortunate troops were surrounded by Prussians and forced to retreat to the Swiss border (detail of *Panorama of Lucerne* by the Swiss painter Édourd Castres, 1881). The French forces disarmed and stayed in Switzerland. Another example of the grandeur and the decadence of the Bourbakis...

this way with their science and imagination." Most Bourbakis would be more than happy to repeat Dieudonné's words.

There is no question that the Bourbakian enterprise did have some perverse effects, such as the dogmatic presentation of mathematics and the neglect of the mathematical subjects that didn't interest the members. There is no question that the group and its writings have aged. Nonetheless, there is no doubt that Bourbaki will go down in the history of mathematics. Such an enterprise deserves admiration for its breadth, for its enthusiasm and selflessness, for its strongly collective character. Despite some mistakes, Bourbaki did add a little to "the honor of the human spirit." In an era when sports and money are such great idols of civilization, this is no small virtue.

Acknowledgements

I would particularly like to thank Liliane Beaulieu, who supplied me with copies of her publications about the Bourbaki group, and Michèle Chouchan, who made all the documents in her possession at my disposal and whose radio interviews and book introduced me to the world of Bourbaki. I would also like to thank David Aubin, Arnaud Beauville, Jean-Pierre Bourguignon, Henri and Nicole Cartan, Pierre Cartier, Mireille Chaleyat-Maurel, Catherine Chevalley, Michel Demazure, Catherine Goldstein, Christian Houzel, Jean-Pierre Kahane, Hélène Nocton, Osmo Pekonen, Jean-Paul Pier, Nathalie Rochefort, David Ruelle, Pierre Samuel, Anna Sierpinska, and Sylvie Weil. Their advice and comments, the conversations I had with them, and the documents they contributed are all priceless. Finally, I am grateful to l'Association des Amis de Nicolas Bourbaki for the loan of numerous photographs and to the publisher *Pour la Science* for the confidence and effort it invested in publishing this book.

—*Maurice Mashaal*

Bibliography

This book relies on information from written sources, all listed below, as well as from interviews with witnesses of or participants in Bourbaki's endeavour. Since the format and style of this book does not allow for footnotes, I tried to use other means, direct and indirect, of stating the sources of the information and quotations presented throughout the book. This, however, was impossible to do in a precise and less cumbersome way for the general information about the history of Bourbaki. The two main sources of this basic information are the theses by Judith Friedman (1977) and Liliane Beaulieu (1990). In the case of more specific information, I often explicitly cite the source along with the corresponding information or quotation; if not, I mention the source either explicitly or implicitly near the information or quotation, generally in a preceding paragraph. Except in any case of unfortunate oversight on my part, quotations appearing without a stated source are from interviews I had with the people in question.

—The Author

General References

- Beaulieu, Liliane. *Bourbaki. Une histoire du groupe de mathématiciens français et de ses travaux (1934-1944)*, doctoral thesis (2 vols.). University of Montreal, 1990.
- _____. *La Tribu N. Bourbaki (1934-1956).* Springer, forthcoming.
- Bourbaki, Nicolas. *L'architecture des Mathématiques.* In François Le Lionnais, *Les grands courants de la pensée mathématique.* Éditions des Cahiers du Sud, 1948 (reprint Rivages, 1986).
- Cartan, Henri. *Nicolas Bourbaki and contemporary mathematics*, The Mathematical Intelligencer **2**, pp. 175–180, 1980 (text of a lecture given in Germany on January 8, 1958).
- Chouchan, Michèle. *Nicolas Bourbaki - Faits et légendes.* Éditions du Choix, 1995.
- Dieudonné, Jean. *Mathématiques vides et mathématiques significatives*, in *Penser les mathématiques.* Seuil, 1982.

- ____. *Regards sur Bourbaki.* Analele Universitatii Bucuresti—Mathematica-Mecanica **18** (2), pp. 13–25, 1969 (text of an address given at the Romanian Insititute of Mathematics in October 1968). English translation: *The work of Nicholas Bourbaki.* American Mathematical Monthly **77**, pp. 134–145, 1970.
- Friedman, Judith. *L'origine et le dévelopement de Bourbaki,* thesis for l'École Pratique des Hautes Études en Sciences Sociales, Paris, 1977.

Portraits, Interviews, Testimonials, Biographies

- *André Weil (1906–1998).* Gazette des Mathématiciens, supplement to no. 80, April 1999.
- André Weil (articles about). Notices of the AMS **46**(4), April 1999.
- Andler, Martin. *Entretien avec trois membres de Nicolas Bourbaki.* Gazette des mathématiciens, January 1988.
- Chouchan, Michèle. *Profil perdu. Bourbaki* (radio broadcast). France-culture, Paris, 1990.
- Borel, Armand. *Twenty-five years with Nicolas Bourbaki, 1949–1973.* Notices of the AMS **45**(3), pp. 373–380, March 1998.
- Demazure, Michel and Martin Andler. *Entretien avec André Weil.* Gazette des mathématiciens, October 1991.
- Dugac, Pierre. *Jean Dieudonné, mathématicien complet.* Gabay, 1995.
- Grothendieck, Alexandre. *Récoltes et semailles.* Université des Sciences et Techniques du Languedoc et CNRS, 1985.
- Guedj, Denis. *Nicolas Bourbaki, mathématicien collectif - interview avec Claude Chevalley.* Dédales, November 1981.
- Jackson, Allyn. *Interview with Henri Cartan.* Notices of the AMS **46**(7), pp. 782–788, August 1999.
- Mandelbrot, Benoît. *Chaos, Bourbaki and Poincaré.* The Mathematical Intelligencer **11**(3), pp. 10–12, 1989.
- Roubaud, Jacques. *Mathématique: (récit).* Seuil, 1997.
- Schmidt, Marian. *Hommes de science: 28 portraits.* Hermann, 1990. (Contains portraits of Henri Cartan, Gustave Choquet, Pierre Deligne, Jean Dieudonné, Jean Leray, André Lichnerowicz, Laurent Schwartz, Jean-Pierre Serre, and René Thom).
- Schwartz, Laurent. *Souvenirs sur Jean Dieudonné.* Pour La Science, June 1994.
- ____. *Un mathématician aux prises avec le siècle.* Odile Jacob, 1997. English translation: *A Mathematician Grappling with His Century.* Trans. Leila Schneps. Birkhaüser, 2001.
- Senechal, Marjorie. *The continuing silence of Bourbaki—An interview with Pierre Carter.* The Mathematical Intelligencer **20** (1), pp. 22–28, 1998.

- Weil, André. *L'avenir des mathématiques.* In François Le Lionnais, *Les grands courants de la pensée mathématique.* Cahiers du Sud, 1948 (reprint Rivages, 1986).
- ____. *Œuvres scientifiques.* Springer, 1979.
- ____. *Souvenirs d'apprentissage.* Birkhaüser, 1991. English translation: *The Apprenticeship of a Mathematician.* Trans. Jennifer Gage. Birkhaüser, 1992.

Historical and Sociological Studies

- Andler, Martin. *Les mathématiques à l'École normale supérieure au XXe siècle: une esquisse.* In J.-F. Sirinelli, *École normale supérieure—Le livre du bicentenaire.* P.U.F., 1994.
- Aubin, David. *The withering immortality of Nicolas Bourbaki: a cultural connector at the confluence of mathematics, structuralism and the Oulipo in France.* Science in Context **10**(2), pp. 297-342, 1997.
- Beaulieu, Liliane. *A Parisian café and ten proto-Bourbaki meetings (1934-1935).* The Mathematical Intelligencer **10**(2), pp. 27-35, 1993.
- ____. *Proofs in expository writing - some examples from Bourbaki's early drafts.* Interchange, 21(2), pp. 33-45, 1990.
- ____. *Questions and answers about Bourbaki's early work. 1934-1944.* In Sasaki et al. (Eds.), *The Intersection of History and Mathematics.* Birkhaüser, 1994.
- ____. *Jeux d'esprit et jeux de mémoire chez N. Bourbaki.* In Pnina G. Abir-Am (Ed.), *La mise en mémoire de la science—Pour une ethnographie historique des rites commémoratifs.* Éditions des Archives Contemporaines, 1998.
- ____. *Bourbaki's art of memory.* Osiris **14**, pp. 219-251, 1999.
- ____. *N. Bourbaki.* An article in *Writing the History of Mathematics—Its Historical Development,* Joseph W. Dauben and Christoph Scriba (Eds.), 730 pp., Birkhauser, 2002.
- Cartier, Pierre. *Notes sur l'histoire et la philosophie des mathématiques I. Vie et mort de Bourbaki, II. La création des noms mathématiques : l'exemple de Bourbaki, III. Le structuralisme en mathématiques : mythe ou réalité?.* Preprints of the I.H.E.S. August 1997/March 1998/April 1998.
- Corry, Leo. *The origins of eternal truth in modern mathematics: Hilbert to Bourbaki and beyond.* Science in Context **12**, pp. 137-183, 1998.
- ____. *Modern Algebra and the Rise of Mathematical Structures.* Birkhaüser, 1996.
- Dahan-Dalmedico, Amy. *Les mathématiques au XXe siècle.* In J. Krige and D. Pestre (Eds.), *Science in the twentieth century.* Harwood, 1997.

- Weintraub, E. Roy and Philip Mirowski. *The pure and the applied: bourbakism comes to mathematical economics.* Science in Context **7**(2), pp. 245-272, 1994.

The New Math Revolution

- *La réforme des mathématiques modernes.* Gazette des mathématiciens, no. 54, November 1992.
- Belhoste, Bruno et al (Ed.). *Les sciences au lycée.* Vuibert, 1996.
- Bkouche, Rudolf, Bernard Charlot and Nicolas Rouche. *Faire des mathématiques: le plaisir des sens.* Arnaud Colin, 1991.
- Moon, Bob. *The "New Maths" Curriculum Controversy: An International Story.* The Falmer Press, 1986.
- Piaget, Jean et al. *L'enseignement des mathématiques.* Delachaux et Niestlé, 1955.

Photo Credits

Chapter 1

p. 4 Alsace infos
p. 5 Library of the École Normale Supériuere
p. 7 Hélène Lamicq
p. 8 Roger Viollet
p. 9 Archives of the Association N. Bourbaki
p. 10 Archives of the Association N. Bourbaki
p. 11 [top] Collection of Catherine Chevalley
p. 12 Archives of the Association N. Bourbaki
p. 13 Association N. Bourbaki
p. 14 Anne Nevanlinna
p. 15 Collection of Sylvie Weil
p. 16 Archives of the Association N. Bourbaki
p. 21 Archives of the Association N. Bourbaki

Chapter 2

p. 22 [left] Patrick Weill
p. 22 [right] Pour la Science
p. 24 Henri Cartan
p. 25 Collection of Sylvie Weil
p. 27 E. Pirou
p. 29 [top] Library of Paris VI
p. 29 [bottom] *The Mathematical Intelligencer*
p. 31 Archives of the Association N. Bourbaki
p. 34 ENS
p. 35 Collection of Sylvie Weil
p. 36 Collection of Sylvie Weil
p. 37 Collection of Sylvie Weil

Chapter 3

p. 38 Library of Mathematics Education at Jussieu
p. 45 *Scientific American*
p. 46 [top & bottom] Pour la Science
p. 47 [bottom] Archives of the Academy of Sciences, Paris

p. 48 [top] Pour la Science
p. 48 [bottom] Collection of Catherine Chevalley
p. 49 Roger Viollet

Chapter 4
p. 58 *The Mathematical Intelligencer*
p. 69 Archives of the Association N. Bourbaki

Chapter 5
p. 70 Cliché CSF - René Bouillot
p. 71 DR
p. 77 Collection of Catherine Chevalley
p. 79 [top] Private collection of Hélène Nocton
p. 80 photo of IHES Cliché CSF - René Bouillot
p. 83 Archives of the Association N. Bourbaki

Chapter 6
p. 87 Private collection of M. Chevalley

Chapter 7
p. 97 Maurice Mashaal
p. 98 Maurice Mashaal
p. 99 Maurice Mashaal
p.100 [bottom] *The Mathematical Intelligencer*
p.101 IHES
p.102 DR
p.103 DR

Chapter 8
p. 107 Private collection of Hélène Nocton
p. 109 Archives of the Association N. Bourbaki
p. 111 Archives of the Association N. Bourbaki
p. 113 Private collection of Hélène Nocton

Chapter 9
p. 117 Tate Gallery, London
p. 118 [top] Roger Viollet
p. 121 [top] Private collection of Hélène Nocton
p. 121 [bottom] *The Mathematical Intelligencer*
p. 122 Archives of the Association N. Bourbaki
p. 126 [top] IHES
p. 126 [bottom] PLS-DR
p. 127 IHES - Cliché René Bouillot

p. 129 PLS-DR
p. 131 Archives of the Association N. Bourbaki
p. 133 Archives of the Association N. Bourbaki

Chapter 10
p. 135 PLS
p. 136 Springer-Verlag
p. 137 L.Monier / Gamma
p. 138 Doisneau-Rapho
p. 139 Archives of the Association N. Bourbaki
p. 143 Private collection of Hélène Nocton

Chapter 11
p. 147 *Le Monde* and *Libération*
p. 149 Cliché CSF - René Bouillet
p. 151 Municipal Library of Pau
p. 153 *Panorama de Lucerne*

Appendix

English translations

Chapter 2, p. 26
An extract from Nicolas Bourbaki's artcle in *Comptes Rendus de l'Académie des Sciences*

Measure Theory. — *On a theorem of Carathéodory and measures in topological spaces.* Article[1] by Nicolas Bourbaki, presented by Élie Cartan.

The modern theory of measure and integration[2] leads us to use the word *measure* for any function μ satisfying the following axioms:

1. Each set E of points in the space corresponds to a number μE such that $0 \leq \mu E \leq +\infty$. If E is empty, then $\mu E = 0$.
2. If E is contained in a finite or countable union of sets E_n, then

$$\mu E \leq \sum_n \mu E_n.$$

[1] *session of November 18, 1935.*

[2] *See for example* S. Saks Théorie d l'integrale (Monog. Matem. **3**, *Varsovie, 1933). Cf. de Possel.* C. R. du Séminaire mathématique de M. Julia **2**, *1934-1935. pp. 1-19 and Comptes rendus* **201**, *1935, p. 579.*

Chapter 2, p. 28
Bourbaki's autograph

Shall I compose a long poem? Or prove a theorem? What I have experienced is all too wonderful. One word remains: *Merci*

N. Bourbaki 7/21/39
(Translated from Poldevian by A. Weil)
One of Bourbaki's wives
Eveline Weil

Chapter 4, p. 53
An extract from Issue 29 of *La Tribu*

Extract from Issue 29 of *La Tribu*
(Conference of the Incarnation of the Trotting Donkey)

To the tune of *En descendent la rue d'Alger*

The instructions for this treatise (x2)
Are the simplest and the neatest (x2)
If something isn't clear
Oh well
Don't worry, never fear!
Abstraction'll ring the bell. (x2)

Alphabets from all the nations (x2)
Will be used in explanations (x2)
For clarity's sake, we want
As usual
To use the tiniest font
For anything useful. (x2)

The exercises enticing (x2)
Their statements so inviting, (x2)
But actually two-thirds
(Ahem)
Are false or for the birds,
So don't waste time on them. (x2)

Our notation, as you'll see (x2)
Is improved as it can be (x2)
And the measure of its worth
We sing
Is that no one on Earth
Will understand a thing. (x2)

The order of exposition (x2)
Is crafted with great erudition (x2)
We put the trivial fluff
In front!
For all the useful stuff
In cor'll'ries you must hunt. (x2)

Chapter 4, p. 54
An extract for the instructions for using Bourbaki's treatise

[...]

4. To obviate this difficulty to some extent, examples referring to facts the reader may know from elsewhere, but which have not already been discussed in the treatise, appear throughout the text. Such examples are also placed between two asterisks: * . . . *. Surely most readers will find that these examples facilitate their understanding of the material and would prefer not to skip them, even in the first reading; however, skipping them would certainly not pose any problems from a logical point of view.

5. The reader who would like to form a general idea of the book's content before starting to study it in detail should read the summary of results appended, in principle, to each volume. These summaries are also intended for impatient readers who would like to begin studying specialized problems as quickly as possible. To the greatest extent possible, these summaries contain most of what is necessary for the following sections.

6. The logical framework of each chapter consists of the *definitions*, *axioms*, and *theorems* appearing in that chapter. These are of vital importance for the material presented later in the book. The less important results, or the ones that follow easily from the theorems, appear as "propositions," "lemmas," "corollaries," "remarks," and so forth, with those that can be omitted in the first reading printed in small characters. Particularly important comments about a theorem are found under the heading "scholium."

7. Some passages have the purpose of warning the reader about serious errors he might make. These passages are marked with the symbol ☡ ("dangerous curves ahead").

Chapter 4, p. 55
The dedication to Enrique Freymann

Enrique Freymann, our editor and friend, passed away while the proofs of this volume were being corrected. Although he had published sixteen booklets of our work, he would not have the satisfaction of seeing the first Book completed.

It was almost twenty years ago that Enrique Freymann, ignoring advice he received from all directions, decided to welcome a completely unknown author.

His confidence helped us greatly throughout our enterprise. From the very beginning, he never stopped following our work with an understanding for which we are grateful.

Today, in honor of a friend who brought art to his job as an editor, we dedicate this volume

TO THE MEMORY
OF
ENRIQUE FREYMANN

Chapter 4, p. 58
Some exercises from the book on topological vector spaces

Exercises.—1) Give an example of a non-compact, closed subset of $\mathbf{R}^2$ whose convex hull is not closed.

2) Show that the convex hull of a compact subset of $\mathbf{R}^n$ is compact (see §1, exer. 8).

¶3) Let I be the interval $(0, 1)$ of $\mathbf{R}$, and let F be the vector space $C(I)$ of continuous numerical functions on I. Let E be the product space $\mathbf{R}^F$ (the set of finite numerical functions defined on F). For each $a \in I$, let ϵ_a be the element of E satisfying $\epsilon_a(f) = f(a)$ for every $f \in F$.

a) Show that the set K containing ϵ_x for all x in I is compact in E.

b) Let μ be an element of E such that $\mu(f) = \int_0^1 f(t)dt$ for any $f \in F$ (the "Lebesgue measure"). Show that μ is a limit point of the convex hull of K in E but does not lie in this convex hull itself. (See Functions of One Real Variable. chap. II, §1, prop. 5.)

Chapter 4, p. 59
An extract from the book on algebra

6. Properties of Ideals

In this section and the following, only left ideals will be considered; we leave it to the reader to formulate the corresponding propositions for right ideals and bilateral ideals.

Let A be a ring, $\mathfrak{a}$ be a left ideal of A, and B be a subring of A. The intersection $B \cap \mathfrak{a}$ is a left ideal in the ring B. In particular, if $\mathfrak{a} \subset B$, then $\mathfrak{a}$ is a left ideal in B. Conversely, however, a left ideal *in* B is not necessarily a left ideal *in* A.

As left ideals of A are just subgroups stable under the group actions on A, all the properties of subgroups stable under group actions (§6, no. 10) apply to left ideals. Thus the intersection of a family $(\mathfrak{a}_i)$ of left ideals is a left ideal. The family of left ideals containing a given subset M of A contains a smallest ideal; this is called the left ideal generated by M, and M is called the set of generators of this ideal.

Chapter 8, p. 105
An extract from the book on general topology

The *distance* between two subsets A and B of E is the number $d(A, B) = \inf_{x \in A, y \in B} d(x, y)$. When B contains a single point x, we write that $d(x, A)$ in place of $d(\{x\}, A)$, and this number is called the *distance from the point* x *to the set* A. The function $x \to d(x, A)$ is uniformly continuous in E. For all $\alpha > 0$, the set $V_\alpha(A)$ (resp. $W_\alpha(A)$) is an open neighborhood (resp. closed neighborhood) of A. The closure of the set A is the set of $x \in E$ such that $d(x, A) = 0$.

In particular, for any $x \in E$, the set $V_\alpha(x)$ (resp. $W_\alpha(x)$) is called an *open ball* (resp. *closed ball*) of *center* x and *radius* α and is an open set (resp. closed set). The set of $y \in E$ such that $d(x, y) = \alpha$ is called the *sphere* of center x and radius α and is a closed set (which could be empty).

Chapter 8, p. 108
An extract from the book on topological vector spaces

Barreled Spaces

1. In a locally convex space E, a set that is closed, convex, balanced, and absorbing is called a *barrel.* A *barrel* absorbs all bounded, complete, convex subsets (chap. III, §3, lemma 1). When E is quasi-complete, a barrel in E absorbs all bounded subsets of E.

Chapter 8, p. 112
A poem by Pierre Samuel

A Pastiche of Mallarmé's "Le Cygne" by Pierre Samuel (1945)

The Filter
O mighty, o formal, o lucid Bourbaki,
Won't you deliver us from our state of crisis,
From longwinded Goursat, the mirror of Analysis,
That backwards defender of a past that was freed?

The sequences of yesteryear found themselves unbounded,
But useless, and mindlessly used
By the clumsy student, he who Valiron confused
In gloomy classes leaving poor souls confounded.

Not knowing the secrets of topologies
Imposed on a space, nor of you or your studies,
He floats in his language so full of mysteries.

He contemplates, as if drugged by a philtre,
The closure, a cloak he can't understand,
Draped over a compact set by the FILTER.

Chapter 8, p. 113
A sonnet by André Weil

A sonnet composed by André Weil at the Chançay conference (September 1937)

Behold a collection of vectors.
A field acts alone, abstract, commutative.
The dual remains afar, solitary and plaintive,
Searching for the isomorphism but finding it elusive.

Suddenly, bilinear, a spark flew,
Begetting the twice-distributive operator.
In the product's net, all the captive vectors
Will endlessly celebrate the more beautiful structure.

But the basis disturbed this aeolian hymn:
The exultant vectors have coordinates.
Cartan knows not what to do and understands not a thing.

And it's the end. Operators, vectors, ruined.
A wretched matrix expires. The naked field
Turns in on itself, embraced in its self-imposed laws.

Chapter 9, p. 128
A poem by Arnold Seiken

One example of the animosity towards Bourbaki is this double poem of questionable taste that the American mathematician Arnold Seiken wrote about French mathematics.

A Double Dactylic Overview of French Mathematics:
Sublime and Other

Sublime

Higgledy-piggledy
J. Henri Poincaré
Elegant theorems
Proved without fuss,

Classified manifolds
Diffeomorphically
Using duality
Also genus.

Other

Higgledy-piggledy
Nicholas Bourbaki
Rose from the dead,
Mathematics to save,

Churned out dull treatises
Hyperkinetically,
Proving that he should have
Stayed in the grave.

The Mathematical Intelligencer, 1995.

Chapter 10, p. 140
The definition of complex numbers in terms of matrices, as taught to twelfth-grade students.

2.1 THE FIELD $\mathbb{C}$ OF MATRICES $\begin{pmatrix} a & -b \\ b & a \end{pmatrix}$

2.1.1 Definition.

Let $\mathcal{M}$ be the set of square $2x2$ matrices with real entries. Over the field $\mathbb{R}$ of real numbers, $\mathbb{C}$ is defined as the subset of matrices $M(a,b)$ of the form:

$$M(a,b) = \begin{pmatrix} a & -b \\ b & a \end{pmatrix},$$

for all a and b.